Modelling with AutoCAD
Release 13 for Windows

Robert McFarlane
MSc, BSc, ARCST, CEng, MIMech E, MIEE, MILog, MIED

Senior Lecturer, Department of Integrated Engineering, Motherwell College

A member of the Hodder Headline Group
LONDON • SYDNEY • AUCKLAND
Copublished in North, Central and South America by
John Wiley & Sons Inc., New York • Toronto

First published in Great Britain 1997 by Arnold,
a member of the Hodder Headline Group,
338 Euston Road, London NW1 3BH

Copublished in North, Central and South America by
John Wiley & Sons Inc., 605 Third Avenue,
New York, NY 10158-0012

© 1997 Robert McFarlane

All rights reserved. No part of this publication may be reproduced or
transmitted in any form or by any means, electronically or mechanically,
including photocopying, recording or any information storage or retrieval
system, without either prior permission in writing from the publisher or a
licence permitting restricted copying. In the United Kingdom such licences
are issued by the Copyright Licensing Agency: 90 Tottenham Court Road,
London W1P 9HE.

Whilst the advice and information in this book is believed to be true
and accurate at the date of going to press, neither the author nor the publisher
can accept any responsibility or liability for any errors or omissions
that may be made.

British Library Cataloguing in Publication Data
A catalogue record for this book is available from the British Library

Library of Congress Cataloging-in-Publication Data
A catalog record for this book is available from the Library of Congress

ISBN 0 340 69251 0
ISBN 0 470 23737 6 (Wiley)

Produced by Gray Publishing, Tunbridge Wells, Kent
Printed and bound in Great Britain by The Bath Press, Bath
and Edinburgh Press Ltd, Edinburgh

Modelling with
AutoCAD

City College Norwich

Other titles from Bob McFarlane

Beginning AutoCAD   ISBN 0 340 58571 4

Progressing with AutoCAD   ISBN 0 340 60173 6

Introducing 3D AutoCAD   ISBN 0 340 61456 0

Solid Modelling with AutoCAD   ISBN 0 340 63204 6

Starting with AutoCAD LT   ISBN 0 340 62543 0

Advancing with AutoCAD LT   ISBN 0 340 64579 2

3D Draughting using AutoCAD   ISBN 0 340 67782 1

Beginning AutoCAD R13 for Windows   ISBN 0 340 64572 5

Advancing with AutoCAD R13 for Windows   ISBN 0 340 69187 5

Contents

	Preface	vii
Chapter 1	Extruded three-dimensional models	1
Chapter 2	Three-dimensional coordinate systems and input	10
Chapter 3	Wire-frame modelling	18
Chapter 4	Modifying 3D models	34
Chapter 5	Dimensioning 3D models	37
Chapter 6	Hatching in 3D	43
Chapter 7	Viewports	48
Chapter 8	Viewpoint	58
Chapter 9	Centring viewports	68
Chapter 10	Surface modelling	72
Chapter 11	3DFACE, PFACE and 3DPOLY	74
Chapter 12	3DMESH	86
Chapter 13	Ruled surface	88
Chapter 14	Tabulated surface	94
Chapter 15	Revolved surface	98
Chapter 16	Edge surface	102
Chapter 17	Model space and paper space	108
Chapter 18	Three-dimensional multi-view drawings	114
Chapter 19	Three-dimensional geometry	124
Chapter 20	Dynamic viewing	132
Chapter 21	Viewport specific layers	141
Chapter 22	Solid modelling introduction	145
Chapter 23	The basic solid primitives	151
Chapter 24	The swept solid primitives	163
Chapter 25	Boolean operations and composite solids	173
Chapter 26	Composite model 1 – a machine support	178
Chapter 27	Composite model 2 – a backing plate	183
Chapter 28	Composite model 3 – a pipe flange	188
Chapter 29	Modifying solids – fillet and chamfer	191
Chapter 30	Regions	196
Chapter 31	Moving solids	202
Chapter 32	Slicing and sectioning solids	205
Chapter 33	A detailed drawing	213
Chapter 34	Solid model block assembly	223
Chapter 35	Dynamic viewing solid models	229
Chapter 36	A solid model house	231
Chapter 37	Finally	236
	Tutorials	237
	Index	247

Preface

This book is intended for the Release 13 user who wants to learn about modelling. The book's aim is to demonstrate how the user can construct wire-frame models, surface models and solid models and introduces the concept of multiple viewports as well as model and paper space. Release 13 is the most powerful AutoCAD package so far introduced to the market, and the Windows platform (which is the only one considered in this book) is very user friendly.

The book will prove an invaluable aid to a wide variety of users, ranging from the capable to the competent. While it has not been designed with any course in mind, it will assist students who are studying City & Guild courses, as well as those BTEC and SCOTVEC students who require 3D and solid modelling in their studies. Students at higher institutions will also find it suitable to their needs, and industry users will find it a useful reference book.

Reader requirements

There are four basic requirements which I think are important for using this book:

1. the ability to draw and edit with Release 13
2. a knowledge of icons and toolbars
3. an understanding of how to use dialogue boxes
4. the ability to open and save drawings to a named directory.

Using the book

The book is essentially a self-teaching package with the reader working interactively through exercises using information supplied. The prompts and responses are listed in the order they occur, and the exercises are backed up with tutorial material. Dialogue boxes will be shown where appropriate.

The following points are important:

1. The named directory for drawings is **R13MODEL** in the C: drive – this should be made before starting any exercises.
2. Icon selection will be displayed when the required icon is used for the first time.
3. Menu bar selection will be in bold type, e.g. **Draw–Line**.
4. Keyboard entry will also be in bold type, e.g. **VPOINT, UCS**, etc.
5. The symbol **<R>** (or <RETURN>) will require the user to press the enter/return key.

Note

All the worked examples were completed with Release 13, and I have tried to correct any mistakes in the text and drawings. Should errors occur, then I apologize in advance for them, and hope that they do not spoil your learning experience. Modelling with Release 13 is fascinating and should give you satisfaction and enjoyment. Any comments you have about how to improve the material in the book would be greatly appreciated.

Good luck!

Chapter **1**

Extruded three-dimensional models

In this chapter we will investigate one of the first types of three-dimensional (3D) modelling achieved with AutoCAD – the extruded (or $2^1/_2$D) model. The chapter will also introduce the user to some of the essential 3D commands.

An extruded model is drawn upwards (or downwards) from a horizontal plane – called the ELEVation plane. The actual extruded height (or depth) is called the THICKness, and can be positive (+) or negative (−) relative to the set elevation plane. This extruded thickness is **always perpendicular** to the elevation plane. The basic terminology is shown in Fig. 1.1.

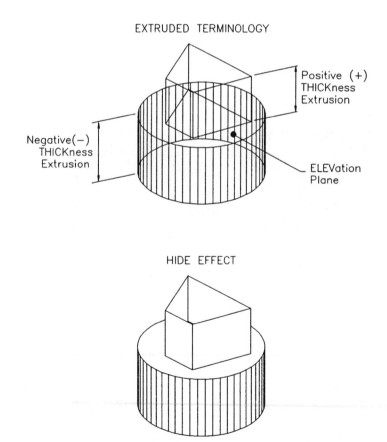

Figure 1.1 Basic terminology.

2 · Modelling with AutoCAD

The elevation and thickness can be 'set':

a) by entering ELEV at the command line
b) from the menu bar with Data–Object Creation ...
c) selecting the Object Creation icon from the Object Properties toolbar.

Options *b*) and *c*) give the Object Creation Modes dialogue box as Fig. 1.2, but we will use all three methods in our examples.

Extruded 3D standard sheet

We will create our extruded 3D models on an A3 standard sheet with the following settings:

- limits: 0,0 to 420,297
- units: metric decimal to two decimal places
- drawing aids: BLIPS off; GRID 10; SNAP 5
- layer: MODEL, continuous, red, current
- save as: **C:\R13MODEL\EXTSTDA3**

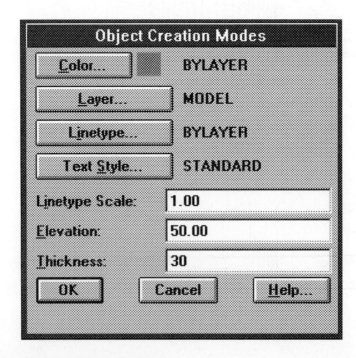

Figure 1.2 Object Creation Modes dialogue box.

Extruded 3D example 1

This worked example is illustrated in Fig. 1.3, so open your EXTSTDA3 standard sheet and follow the instructions given. Suggested toolbars are Draw, Modify and Object Snap.

Step 1 – the first ELEVation

1 At the command line enter **ELEV** <R>
 prompt New current elevation<0.00>
 enter **0** <R>
 prompt New current thickness<0.00>
 enter **50** <R>
 Nothing appears to have happened?

2 Select the LINE icon and draw a square:
 From point **40,40** <R>
 To point **@100,0** <R>
 To point **@100<90** <R>
 To point **@–100<0** <R>
 To point **c** <R>

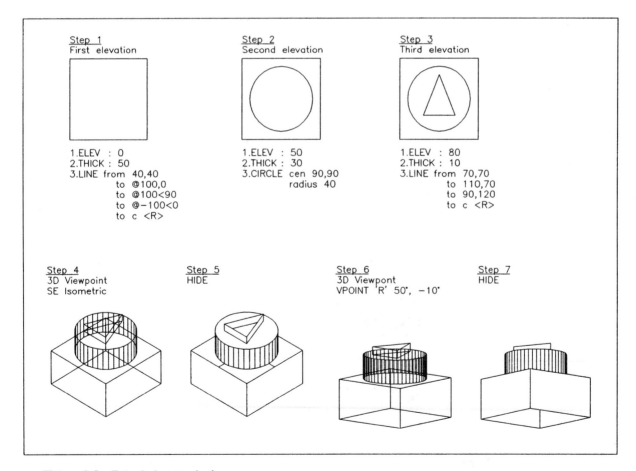

Figure 1.3 Extruded example 1.

4 *Modelling with AutoCAD*

Step 2 – the second elevation

1 From the menu bar select **Data**
                      Object Creation ...
  prompt   Object Creation dialogue box
  with   1. Elevation: 0.00
            2. Thickness: 50.00
  respond 1. alter elevation to **50**
            2. alter thickness to **30** – Fig. 1.2
            3. pick OK

2 Select the CIRCLE icon (Center,Radius) and draw a circle:

  a) center at 90,90
  b) radius 40.

Step 3 – the third elevation

1 Select the Object Creation icon from the Object Properties toolbar and:
  prompt   Object Creation Modes dialogue box
  with   1. Elevation: 50.00
            2. Thickness: 30.00
  respond 1. alter elevation to **80**
            2. alter thickness to **10**
            3. pick OK

2 With the LINE icon draw:
  From point   **70,70** \<R\>
  To point   **110,70** \<R\>
  To point   **90,120** \<R\>
  To point   **c** \<R\>

3 Now have a triangle inside a circle inside a square, and appear to have an ordinary 2D-plan type drawing. Each of the three shapes has been created on a different elevation plane:
  a) square: on elevation 0
  b) circle: on elevation 50
  c) triangle: on elevation 80

Step 4 – viewing in 3D

To 'see' the created extruded model in 3D, it is necessary to use the 3D viewpoint command, so:

1 From the menu bar select **View**
                      3D Viewpoint Presets
                      SE Isometric

2 The model is now displayed in 3D and 'fills the screen'.

3 The orientation of the model may be such that it is difficult to know if you are looking down on the top, or looking up at the bottom. This is common with all 3D modelling and is called **AMBIGUITY**. Another command is required to 'remove' this ambiguity.

4 At this stage, save your model as **C:\R13MODEL\EXT_1**.

Extruded three-dimensional models **5**

Step 5 – the HIDE command

1 At the command line enter **HIDE** <R>
prompt   Regenerating drawing
              Hiding lines: 100% done

2 The model should now be easier to 'see' as it is displayed with all the hidden lines removed. Hidden lines are those which are at 'the back of the object' as you look at it. Think about the power to do this?

Step 6 – another viewpoint

The model is being viewed from above, but it is possible to change our viewpoint. We will be investigating this command in more detail in a later chapter, but at present:

1 At the command line enter **VPOINT** <R>
prompt   Rotate/<View ...
enter   **R** <R> – the rotate option
prompt   Enter angle in *XY*-plane from *X*-axis
enter   **50** <R>
prompt   Enter angle from *XY*-plane
enter   **−10** <R>

2 The model is displayed from a different viewpoint but is displayed without hidden line removal.

Step 7 – the HIDE command

1 At the command line enter **HIDE** <R>

2 The new viewpoint model is displayed with hidden line removal.

3 Display the model without hidden line removal by entering **REGEN** <R> at the command line.

This completes the first extruded model example.

Notes

1 Extruded models have no top or bottom surfaces, only sides. This should be apparent when hide is on.
2 With the ERASE icon pick any one of the 'base' lines, and a complete 'side' is erased. Undo this erase command.
3 Using the ERASE icon again, pick any point on the top 'circle' and the complete 'cylinder' is erased. Undo again.
4 The HIDE command is essential with 3D modelling.
5 REGEN removes the effect of HIDE, and restores the model to its original display.

6 Modelling with AutoCAD

Extruded 3D example 2

This model is created in the same manner as the first example, i.e. by setting the elevation and thickness. I have tried to make this example more interesting, so open your extruded EXTSTDA3 standard sheet and refer to Fig. 1.4.

Step 1 – the base

1 Using the Object Creation Modes dialogue box set:
Elevation: 0
Thickness: 30

2 With the POLYLINE icon from the Draw toolbar, draw a polyline of width 0 :
From point   **50,50** <R>
To point   **@100,0** <R>
To point   **@0,100** <R>
To point   **@–100,0** <R>
To point   **c** <R>

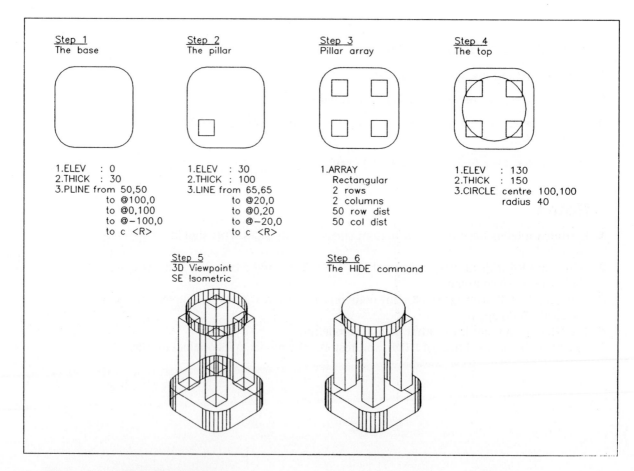

Figure 1.4 Extruded example 2.

3 From the menu bar select **Construct**
Fillet

prompt Polyline/Radius ...
enter **R** <R> – the radius option
prompt Enter fillet radius
enter **20** <R>

4 Select **Construct–Fillet** again and:
prompt Polyline/Radius ...
enter **P** <R> – the polyline option
prompt Select 2D polyline
respond **pick any point on the polyline**

5 The drawn polyline is filleted at the four corners.

Step 2 – the first pillar

1 Using the Object Creation Modes dialogue box, set:
- Elevation: 30
- Thickness: 100

2 With the LINE icon draw a 20-unit square, the lower left corner being at the point 65,65.

Step 3 – arraying the pillar

1 From the Modify toolbar, select the RECTANGULAR ARRAY icon and:
prompt Select objects
respond **window the square** then right-click
prompt Number of rows and enter **2** <R>
prompt Number of columns and enter **2** <R>
prompt Unit cell or row ... and enter **50** <R>
prompt Distance between columns and enter **50** <R>

2 The square is arrayed in a 2 × 2 pattern.

Step 4 – the top

1 At the command line enter **ELEV** <R> and:
prompt New current elevation<30.00>
enter **130** <R>
prompt New current thickness<100.00>
enter **15** <R>

2 Draw a circle, centre at the point 100,100 with a 40 radius.

Step 5 – the 3D viewpoint

1 From the menu bar select **View**
3D Viewpoint Presets
SE Isometric

2 The model is displayed in 3D, but is 'cluttered'.

8 *Modelling with AutoCAD*

Step 6 – hiding the model

1 From the menu bar select **Tools**
                                    Hide

2 The model is displayed with hidden line removal, and is quite interesting?

Step 7 – saving the model

1 From the menu bar select **File–Save As ...** and:
prompt   Save Drawing As dialogue box
enter   **\R13MODEL\EXT_2** as the file name

Task

Before leaving this exercise, I want to introduce an additional command which should make you appreciate what 3D modelling is all about.

1 Still have the example 2 model on the screen?

2 From the menu bar select **View**
                                3D Viewpoint Presets
                                Plan View
                                World

3 The model is displayed 'as drawn'.

4 Using the PROPERTIES icon from the Object Properties toolbar, change the colour of the following entities:
a) the base: red – should be?
b) the four pillars: blue
c) the circular top: green.

5 Restore the SE Isometric viewpoint – easy?

6 From the menu bar select **Tools**
                                Shade
                                16 Color Hidden Line
prompt   interesting model display?

7 From the menu bar select **Tools**
                                Shade
                                16 Color Filled
prompt   very impressive model display?

8 That completes the additional exercise and you can now proceed to the activity included in this chapter.

Summary

1 An extruded model is created from an ELEVation and a THICKness.
2 Extruded models have no top or bottom surfaces – only sides.
3 The elevation and thickness can be 'set' from the command prompt line or from the Object Creation Modes dialogue box.
4 Models are viewed in 3D with the VIEWPOINT command.
5 The HIDE command displays a 3D model with hidden line removal.
6 The REGEN (regenerate) command restores the model with all lines. This command must be entered from the keyboard.
7 The SHADE command gives useful displays with coloured entities.

Activity

Creating extruded 3D drawings is fairly simple, and I have included one activity for you to attempt.

Tutorial 1: Using the reference sizes given, create an extruded 3D model of the half-coupling using your EXTSTDA3 standard sheet. All the elevation and thickness values are on the drawing, but to help you:
a) Base: Elevation 0; Thickness 40
b) Bolt: Elevation 40; Thickness 15
c) Shaft: Elevation 40; Thickness 20.

When the model has been drawn, try the following viewpoints:
a) 3D Viewpoint Presets SE Isometric from the menu bar
b) Enter VPOINT <R> at the command line then:
    prompt angle 1, enter 315
    prompt angle 2. enter −5
c) Remember HIDE
d) SHADE?
e) Try some other viewpoints.

Chapter 2

Three-dimensional coordinate systems and input

AutoCAD uses two coordinate systems:

1. the world coordinate system (**WCS**) and
2. the user coordinate system (**UCS**).

We will discuss each system before considering how coordinate input is entered.

The WCS

All readers should be familiar with the basic 2D coordinate concept of a point described as P1 (30,40) – Fig. 2.1. Such a point has 30 units in the positive X-direction, and 40 units in the positive Y-direction. These coordinates are relative to an XY-axes system with the origin at the point 0,0 and for 2D draughting this origin is normally positioned at the lower left-corner of the screen. Such a system is satisfactory for 2D draughting but is not sufficient for 3D draughting.

Drawing 3D models requires a third axis (the Z-axis) to enable three-dimensional co-ordinates to be used. The screen monitor is a flat surface, and so it is difficult to show a three-axis coordinate system on it. AutoCAD overcomes this difficulty by using an **ICON**. In 3D this icon can be moved to different points on the screen and can be orientated on (or about) objects.

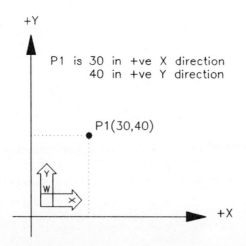

Figure 2.1 Two-dimensional coordinate entry with the WCS at origin (0,0).

Figure 2.2 shows the basic idea of how the icon has been constructed. The X- and Y-axes are shown in their correct orientations, and the Z-axis is pointing outwards to the user. The **W** on the icon indicates that the user is viewing in the world coordinate system. The origin is the point 0,0,0 and is positioned at the lower left-corner of the screen – as it is in 2D. You will find out later that this origin point can be moved to suit the model being created.

Thus the point P2 (30,40,50) has 30 units in the positive X-direction, 40 units in the positive Y-direction and 50 units in the positive Z-direction. Similarly the point P3 (–40,–50,–30) has 40 units in the negative X-direction, 50 units in the negative Y-direction and 30 units in the negative Z-direction.

In the previous chapter on extruded models, we created the model while in the WCS.

The UCS

The user coordinate system allows the user to:

1 Move the origin to any point (or object) on the screen.
2 Align the UCS icon to suit any 'plane' on the component being created.
3 Rotate the icon about the X-, Y- and Z-axes.
4 Save UCS positions for future recall.

The appearance of the UCS icon alters depending on:

a) its orientation, i.e. how it is 'attached' to a component
b) the viewpoint entered.

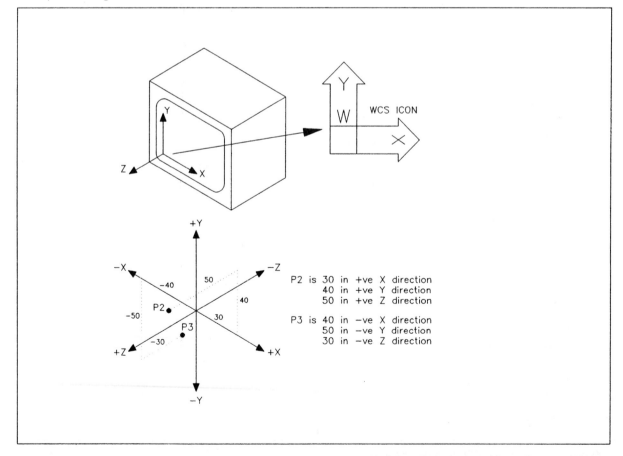

Figure 2.2 Three-dimensional coordinate input.

12 Modelling with AutoCAD

UCS icon exercise

As an introduction to the UCS icon, the following sequence of operations will be used. I should explain that several of the commands will be new to some readers, but they will be explained later. The object of this exercise is to make you aware of the 'versatility' of the UCS icon.

1. From the menu bar select **File–New** and pick OK from the Create New Drawing dialogue box. Refer to Fig. 2.3.

2. The icon displayed at the lower left corner of the screen has a W on it, indicating that we are working with the WCS – fig. (a).

3. From the menu bar select **Options**
   UCS
   Icon, i.e. remove tick – Icon OFF
   prompt No icon displayed?

4. From the menu bar select **Options**
   UCS
   Icon, i.e. add tick – Icon ON
   prompt Icon is displayed?

5. Select from the menu bar **Options–UCS–Icon origin**
   prompt No change in icon?

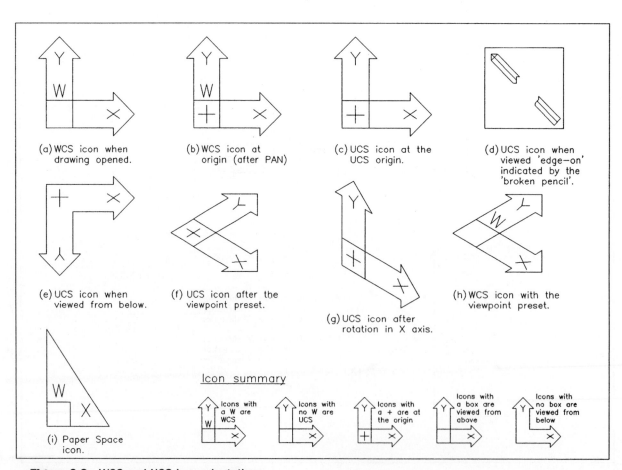

Figure 2.3 WCS and UCS icon orientations.

6 At the command line enter **PAN** <R>
prompt Displacement and enter **0,0** <R>
prompt Second point and enter **50,50** <R>
prompt The icon will be displayed as fig. (b) with a + sign added to the 'box'.

7 With the SNAP on, move the cursor cross-hairs onto the icon + and note the coordinate display in the status bar – it should read 0.0000,0.0000.

8 Now PAN from 50,50 to 0,0 and the icon is again displayed as fig. (a). This icon is at the point (0,0), but as the point (0,0) is at the extreme corner of the screen, AutoCAD cannot position the icon on it. The icon is 'placed' at the side – hence the reason for no + in the box.

9 At the command line enter **UCS** <R> and:
prompt Origin/ZAxis ...
enter **O** <R> – the origin option
prompt Origin point<0,0,0>
enter **100,100** <R>
prompt The icon moves to the entered point and is displayed as fig. (c). Move the cursor onto the + and the coordinate display in the status bar will be 0.0000,0.0000.
In this position:
a) there is no W, i.e. it is a UCS icon
b) there is a +, i.e. icon is at the origin

10 At the command line enter **UCS** <R>
prompt Origin/ZAxis ...
enter **X** <R> – the *X*-axis option
prompt Rotation angle about *X*-axis
enter **90** <R>
prompt The icon is displayed as a 'broken pencil' – fig. (d). This means that we are looking at the icon 'edge on'.

11 Enter **UCS** <R> at the command line and:
prompt Origin/ZAxis ...
enter **X** <R>
prompt Rotation angle about *X*-axis
enter **90** <R>
prompt The icon is displayed as fig. (e), i.e. we are viewing it from underneath. The + is still there, so the icon is still at the origin point.

12 Now enter **UCS-X-180** and the icon should be displayed as fig. (c).

13 From the menu bar select **View**
                        3D Viewpoint Presets
                        SE Isometric
prompt The icon is displayed 'in 3D' as fig. (f). It is still at the origin (+) and is a UCS icon (no W).

14 At the command line enter **UCS-X-90** and the icon will be displayed as fig. (g).

15 Undo this *X*-axis rotation with **U** <R> or by selecting the undo icon from the Standard toolbar.

16 From the Standard toolbar, select the UCS flyout and pick the World Icon
prompt Icon displayed as fig. (h) and is still in the SE Isometric orientation, but no +?

14 Modelling with AutoCAD

17 From the menu bar select **View**
3D Viewpoint Presets
Plan View
World

prompt Icon back to fig. (a) display at the start of this exercise.

18 Double left-click on the word MODEL in the status bar and the icon will change to fig. (i). This is the Paper Space icon which we will discuss in a later chapter.

19 This completes this introduction to the icon, and in Fig. 2.3 I have summarized the main icon points which are worth remembering.

Orientation of the UCS

The exercise which has been completed has shown that the UCS icon can be moved to different positions and rotated about the three axes. It is thus important for the user to be able to determine the correct orientation of the icon, i.e. how the X-, Y- and Z-axes will 'lie' in a new icon position.

The axes orientation is determined by the right-hand rule and this is demonstrated in Fig. 2.4. The knuckle of the hand is at the origin and the position of the thumb, index finger and second finger determine the positions of the X-, Y- and Z-axes, respectively.

Three-dimensional coordinate input

To draw accurately coordinate input is required. With 3D draughting there are three types of coordinate available each having both absolute or relative modes of entry. The three input types are:

Type	Absolute	Relative
a) Cartesian:	10,20,30	@30,20,10
b) Cylindrical:	20<30,50	@50<20,30
c) Spherical:	40<50<60	@60<40<50

To investigate the different types of coordinate input we will draw some line segments using each entry and also investigate the effect of different icon positions on the input. Once again I should stress that some of the commands used will be new, but all will be explained in a later chapter.

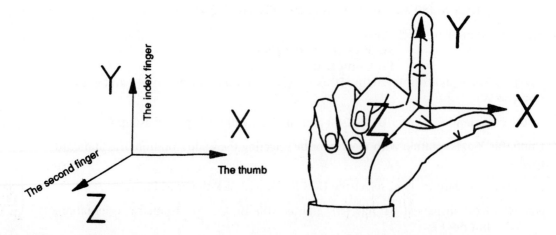

Figure 2.4 The right hand rule.

1 Start AutoCAD and refer to Fig. 2.5.

2 From the menu bar select **Options–UCS** and check that both the Icon and Icon Origin are on, i.e. both have ticks at their names.

3 From the menu bar select **View–3D Viewpoint Presets–SE Isometric** to view the drawing in 3D.

4 With layer 0 (black) current use the LINE icon to draw: from 50,50; to @ 150,0; to @0,80; to @-150,0; to c <R>

5 Make the following new layers:

line1 – red; line2 – green; line3 – blue; line4 – magenta; line5 – yellow.

6 With layer line1 (red) current, use the line icon to draw
From point   **0,0,0** <R>
To point   **@70,80,90** <R>   (relative absolute)   – line 1W
To point   **@70<120,50** <R>   (relative cylindrical) – line 2W
To point   **@50<40<30** <R>   (relative spherical)   – line 3W
To point   right-click

7 Now ZOOM All and PAN from 0,0 to 20,20. The WCS icon is 'attached' to the end of line 1W as Fig. 2.5.

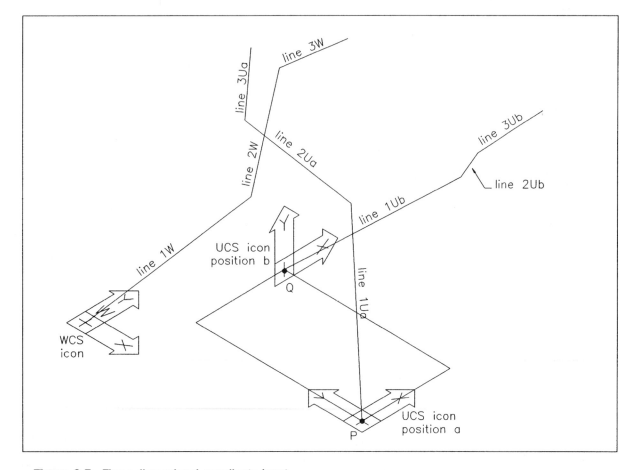

Figure 2.5 Three-dimensional coordinate input.

16 Modelling with AutoCAD

8 At the command line enter **UCS** <R>
    prompt   Origin/ZAxis ...
    enter   **O** <R> – the origin option
    prompt   Origin point <0,0,0>
    respond **INTersection and pick pt P**

9 Repeat the **UCS** <R> command line entry and:
    prompt   Origin/ZAxis ...
    enter   **Z** <R> – Z-axis rotation
    prompt   Rotation angle about Z-axis
    enter   **90** <R>
    prompt   icon is aligned at point P – icon position a in Fig. 2.5.

10 Make layer line2 (green) current and with the LINE icon enter the following:
    From point **0,0,0** <R>
    To point   **@70,80,90** <R>   – line 1Ua
    To point   **@70<120,50** <R>   – line 2Ua
    To point   **@50<40<30** <R>   – line 3Ua
    To point   right-click

11 The coordinate input is the same as before but the lines appear different. This is due to the UCS position.

12 Make layer line3 (blue) current and draw lines using:
    From point ***0,0,0** <R> – yes *
    To point   **@*70,80,90** <R>
    To point   **@*70<120,50** <R>
    To point   **@*50<40<30** <R>
    To point   right-click

13 Three blue lines have been drawn identical to the three red lines drawn with the WCS icon, i.e. using the * 'converts' the input to world, irrespective of the UCS position.

14 At the command line enter the following sequence:
    UCS <R>
    O <R> – the origin option
    INTersection of pt Q
    UCS <R>
    X <R> – rotation about X-axis
    90 <R> – the angle

15 The UCS icon will be aligned as icon position b.

16 Make layer line4 (magenta) current and draw lines:
    From point **0,0,0** <R>
    To point   **@70,80,90** <R>   – line 1Ub
    To point   **@70<120,50** <R>   – line 2Ub
    To point   **@50<40<30** <R>   – line 3Ub
    To point   right-click

17 With layer line5 (yellow) current draw the lines:
    From point ***0,0,0** <R>
    To point **@*70,80,90** <R>
    To point **@*70<120,50** <R>
    To point **@*50<40<30** <R>
    To point right-click

Three-dimensional coordinate systems and inputs **17**

18 Three yellow lines will be drawn identical to the red/blue lines already drawn.

19 This completes the exercise on 3D coordinate input. You can save the drawing if you want, but we will not refer to it again.

Summary

1 There are two coordinate systems – the World Coordinate System (**WCS**) and the User Coordinate System (**UCS**). Each system has its own icon.

2 The WCS is a fixed system, the origin being at the point 0,0,0.

3 The WCS icon is standard and does not change.

4 The UCS allows the user to define the origin, either as a point on the screen or referenced to an existing entity.

5 The UCS icon changes 'appearance' dependant on the viewpoint.

6 The UCS icon can be rotated about the three axes, and its current position can be saved.

7 3D coordinate input can be:
   a) Cartesian, e.g. 10,20,30
   b) Cylindrical, e.g. 10<20,30
   c) Spherical, e.g. 10<20<30

8 Both absolute and relative modes of input are possible with 3D coordinates, e.g.
   a) absolute – 120,130,140
   b) relative – @120,130,140

9 3D coordinate input can be relative to the current UCS position or to the WCS, e.g.
   a) 10,20,30 for UCS entry
   b) *10,20,30 for WCS entry.

10 It is recommended that **3D coordinate input is relative to the UCS position**.

Chapter 3

Wire-frame modelling

A wire-frame model can be considered as similar to a component made from a series of different length pieces of wire, fixed at the connecting ends. Hence the name **wire-frame**.

In this chapter we will create a 3D wire-frame model and use the model to:

a) investigate how the UCS can be 'set and saved
b) add 'objects' and text to a the model.
c) investigate the UCS options.

The model will also be saved to allow us to investigate editing, hatching, dimensioning, viewports and viewpoints in 3D.

This chapter is rather long, as it introduces the user to the basic principles of all 3D modelling.

3D standard sheet

As all our modelling will be on A3 paper, we will create a standard sheet (prototype drawing). This standard sheet will be used for much of the future modelling work as well as for your tutorials.

1 Start AutoCAD.

2 Set the following:
a) BLIPS and GRIPS both OFF
b) SNAP: 5; GRID: 10
c) LIMITS: 0,0 to 420,297
d) GLOBAL LINETYPE SCALE: 12
e) UNITS: Decimal to 2DP
         Decimal angles to 1DP
f) Menu bar **Options–UCS**:   Icon: ON, i.e. tick
                             Icon Origin: ON, i.e. tick

g) LAYERS:

Name	LType	Colour	Status
0	continuous	white(black?)	on
MODEL	continuous	red	on and current
OBJECTS	continuous	blue	on
TEXT	continuous	green	on
DIMS	continuous	magenta	on
HATCH	continuous	cyan	on

Other layers may be added during other chapters.

h) Text style: name   – STD3D
               font   – ROMANS
               values   – accept **all** defaults

i) With layer 0 current, use the LINE icon to draw a rectangle from 0,0 to 380,270. This will serve as a 'baseplane' for our models. Remember to make layer MODEL current again.

Wire-frame modelling **19**

j) Menu bar **View–3D Viewpoint Presets–SE Isometric**
k) Toolbars to suit yourself, but generally recommend Draw, Modify and Object Snap.
l) Save the standard sheet as **C:\R13MODEL\STD3D**.

3 There may be additions to this standard sheet in later chapters.

A) Creating the wire-frame model

1 From the menu bar select **File–New** and enter:
a) **C:\R13MODEL\STD3D** as the Prototype Drawing name.
b) **C:\R13MODEL\WORK3D** as the New Drawing name.
c) pick OK.

2 Ensure layer MODEL is current and refer to Fig. 3.1.

The base

3 Select the LINE icon and draw:

From point	**50,50** <R>	pt1
To point	**@150,0** <R>	pt2
To point	**@80<90** <R>	pt3
To point	**@–150<0** <R>	pt5
To point	**c** <R> – fig. (a).	

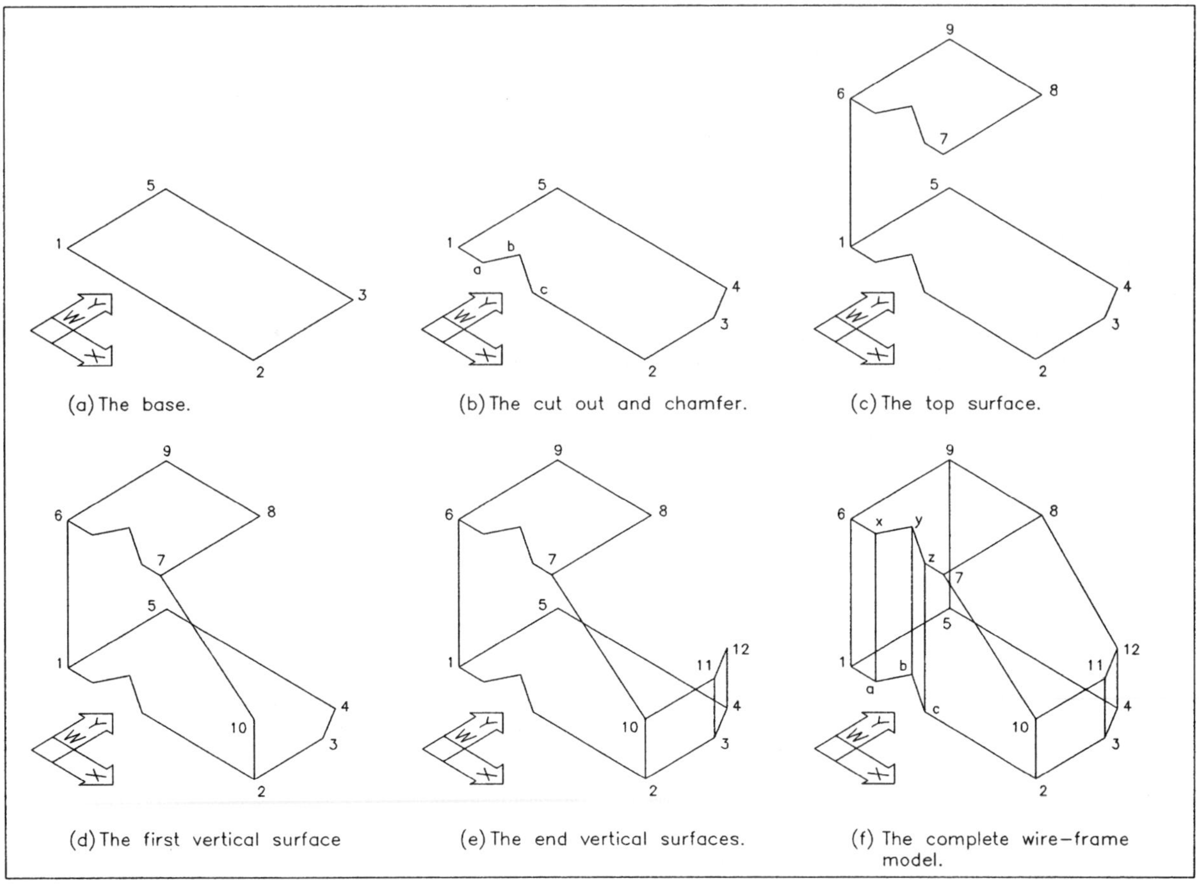

(a) The base. (b) The cut out and chamfer. (c) The top surface.

(d) The first vertical surface (e) The end vertical surfaces. (f) The complete wire-frame model.

Figure 3.1 Construction of the wire-frame model WORK3D.

20 *Modelling with AutoCAD*

4 Add the triangular 'cut-out' using the LINE icon:
From point	**70,50** <R>	pta
To point	**80,70** <R>	ptb
To point	**110,50** <R>	ptc
To point	right-click	

5 TRIM the unwanted line between the lines ab and bc – fig. (b).
Note: when the trim command is used AutoCAD will prompt:
  View is not plan to UCS
  Command results may not be obvious.

6 From the menu bar select **Construct–Chamfer** and:
prompt	Polyline/Distance ...
enter	**D** <R> – the distance option
prompt	Enter first chamfer distance and enter **25** <R>
prompt	Enter second chamfer distance and enter **15** <R>

7 Select **Construct–Chamfer** again and:
prompt	Polyline/Distance ...
respond	**pick line 23**
prompt	Select second line and **pick line 35**

8 This gives the two 'new point' 3 and 4 – fig. (b).

The top surface

9 With the LINE icon draw:
From point	**INTersection and pick pt1**	
To point	**@0,0,100** <R>	pt6
To point	**@75,0,0** <R>	pt7
To point	**@80<90,0** <R>	pt8
To point	**@–75,0,0** <R>	pt9
To point	**@80<–90,0** <R>	pt6
To point	right-click	

10 Need to Zoom-All? – fig. (c).

11 The top surface has been drawn as a rectangle without the cut-out and this will be added
with the copy command. From the Modify toolbar pick the COPY icon and:
prompt	Select objects
respond	**pick lines ab and bc** then right-click
prompt	<Base point or ...
respond	**INTersection and pick pta**
prompt	Second point of displacement
enter	**@0,0,100** <R>

12 Trim the unwanted line on the top surface – fig. (c).

The first vertical surface

13 One line of the 'front' vertical face has already been drawn and we will now add the
remaining lines. With the LINE icon draw:
From point	**INTersection and pick pt2**
To point	**@0,0,40** <R> – pt10
To point	**INTersection and pick pt7**
To point	right-click

14 This gives fig. (d).

Wire-frame modelling **21**

The end vertical surfaces

15 With the COPY icon:
prompt Select objects
respond **pick line 2–10** then right-click
prompt <Base point ...
respond **M** <R> – the multiple option
prompt <Base point ...
respond **INTersection and pick pt2**
prompt Second point ... and **pick pt3** line 3–11
prompt Second point ... and **pick pt4** line 4–12
prompt Second point and right-click

16 With the LINE icon draw:
From point **INTersection and pick pt10**
To point **ENDpoint and pick pt11**
To point **ENDpoint and pick pt12**
To point right-click

17 These lines give fig. (e) and the model is near completion.

Completing the model

18 Use the LINE icon and add lines:
a) from pt5 to pt9
b) from pt12 to pt8
c) from a to x, b to y, c to z.

19 The model is complete – fig. (f) – and you have created your first 3D wire-frame model.

20 From the menu bar select **File–Save** which will update the drawing C:\R13MODEL\ WORK3D named at the start of the exercise. This model will be used for future exercises.

21 Note: the wire-frame model has been constructed with the WCS icon in the same position throughout the exercise, i.e. we have not made any attempt to use the UCS. This is a perfectly valid method of creating 3D models, but problems can occur when objects have to be added to different surfaces on the model when the coordinate input may be difficult to calculate. Using the UCS usually overcomes these types of problem.

22 *Modelling with AutoCAD*

B) Moving around with the UCS

To improve our knowledge of the UCS and how it is used with 3D models we will use the created wire-frame model to add additional entities and text. The sequence is long but worth the time and effort spent. During the sequence of UCS operations, I will use the keyboard entry method to activate the various commands and will discuss the menu bar and toolbar UCS options later in the chapter.

1 Open the wire-frame model **C:\R13MODEL\WORKDRG** or continue from the previous exercise. The icon displayed is WCS. Refer to Fig. 3.2.

2 To enable us to 'see things easier', use the ZOOM-WINDOW option and enter:
First corner: **125,–75** <R>
Other corner: **25,380** <R>

3 At the command line enter **UCS** <R> and:
prompt Origin/ZAxis ...
enter **O** <R> – the origin option
prompt Origin point<0,0,0>
respond **INTersection and pick pt1**

4 The icon 'moves' to point 1 and is a UCS icon – fig. (a). If the icon did not move, select **Options–UCS** from the menu bar and:
a) Icon ON, i.e. tick
b) Icon Origin ON, i.e. tick

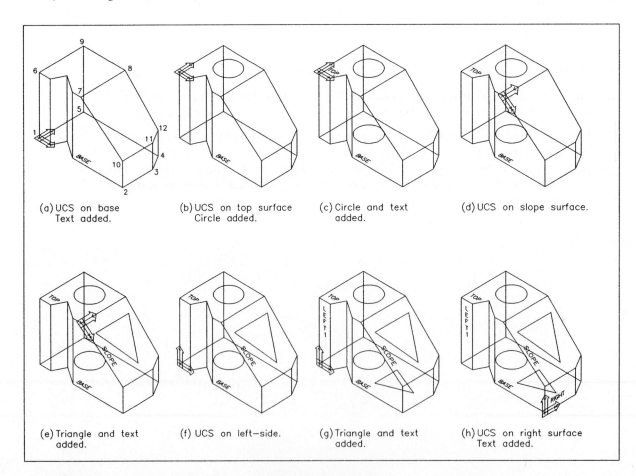

(a) UCS on base Text added.
(b) UCS on top surface Circle added.
(c) Circle and text added.
(d) UCS on slope surface.
(e) Triangle and text added.
(f) UCS on left-side.
(g) Triangle and text added.
(h) UCS on right surface Text added.

Figure 3.2 Investigating the UCS and adding objects and text.

5 Now that the UCS has been defined at point 1, we want to save this 'position' for future recall so at the command line enter **UCS** <R> and:
prompt   Origin/ZAzis ...
enter   **S** <R> – the save option
prompt   ?/Desired UCS name
enter   **BASE** <R>

6 While nothing appears to have happened, the position of the current UCS has been saved with the name BASE.

7 Make layer TEXT (green) current and with the dynamic text icon add the following text item:
a) start point: **60,5**
b) height: 6 and rotation: 0
c) item: **BASE**

8 The model now appears as fig(a).

9 Enter **UCS** <R> at the command line and:
prompt   Origin/ZAxis ... and enter **O** <R>
prompt   Origin point and pick **INTersection of pt6**

10 The icon moves to the top surface and has to be saved, so at the command line enter **UCS** <R>
prompt   Origin/ZAxis ... and enter **S** <R>
prompt   ?/Desired UCS name and enter **TOP** <R>

11 With layer OBJECTS (blue) current use the CIRCLE (Cen,rad) with:
a) centre point: 40,50,0
b) radius: 20

12 A blue circle is drawn on the top surface – fig. (b).

13 Repeat the circle command and enter:
a) centre point: 40,50 −100
b) radius: 20
Can you reason out the centre point coordinates?

14 With layer TEXT current add the following text item:
a) start point: 5,10
b) height and rotation: 6 and 0 respectively
c) item: TOP

15 The model now resembles fig. (c).

16 The UCS has now to be positioned on the slope surface, so enter **UCS** <R> and:
prompt   Origin/ZAxis ...
enter   **3** <R> – the three point option
prompt   Origin point
respond   **INTersection and pick pt7**
prompt   Point on positive portion of the *X*-axis
respond   **INTersection and pick pt10**
prompt   Point on positive-*Y* portion of the UCS *XY*-plane
respond   **INTersection and pick pt8**

17 The UCS moves and is aligned with the sloped surface – fig. (d).

18 At the command line enter **UCS** <R> and save this position as SLOPE.

24 *Modelling with AutoCAD*

19 Text layer still current, so with dynamic text enter:
   a) start point: 40,5
   b) height and rotation: 6 and 0
   c) item: SLOPE

20 With layer OBJECTS current, use the LINE icon to draw:
   From point   **20,10** <R>
   To point   **20,70** <R>
   To point   **@60,–30** <R>
   To point   **c** <R>

21 The blue triangle and green text item are added as fig. (e).

22 At the command line enter **UCS** <R> and:
   prompt   Origin/ZAxis ...
   enter   **R** <R> – the restore option
   prompt   ?/Name of UCS to restore
   enter   **BASE** <R>

23 The UCS is restored to the base position of fig. (a).

24 We now want to rotate the UCS about the *X*-axis and save the new UCS setting, so at the command line enter **UCS** <R> and:
   prompt   Origin/ZAxis ...
   enter   **X** <R> – the rotate about *X*-axis option
   prompt   Rotation angle about *X*-axis
   enter   **90** <R>

25 The UCS is aligned on the left face as fig. (f).

26 Now use the UCS entry again to save the position as LEFT1.

27 With layer OBJECTS current, use the LINE icon and draw:
   From point   **80,0,–20** <R>
   To point   **@0,0,–40** <R>
   To point   **@50,0,20** <R>
   To point   **c** <R>

28 A blue triangle is added to the base surface. You should be able to reason out how this happened with the coordinates entered?

29 With layer TEXT current add the following text item:
   a) centred on: 10,85
   b) height and rotation: 6 and 0
   c) item: L <R>
      E <R>
      F <R>
      T <R>
      1 <R>

30 The model now resembles fig. (g) with these items added.

31 The final UCS position is on the right surface of the model, so at the command line enter **UCS** <R> and:
   prompt   Origin/ZAxis ...
   enter   **O** <R>
   prompt   Origin point
   respond **INTersection and pick pt2**

32 The icon moves to point 2 but is oriented wrongly?

33 Rotate the icon about the Y-axis with the following entries:
 a) UCS <R> – the command
 b) Y <R> – the Y-axis option
 c) 90 <R> – the rotation angle
 d) UCS <R> – that command again
 e) S <R> – the save option
 f) RIGHT <R> – the desired UCS name

34 With layer TEXT current, add the text item RIGHT at the point 10,10 with height 6 and 0 rotation.

35 Your model should resemble fig. (h) and is nearly (but not quite) complete.

Task

The wire-frame model has eleven 'surfaces' and we have set and saved UCS positions for five of these – BASE, TOP, SLOPE, LEFT1 and RIGHT. You now have to set and save the other six UCS positions, i.e. one for each surface, and add an appropriate text item to that surface. I generally make the text item and the UCS name the same as it seems logical?. My suggested names are BACK, REAR, INSHOT1, INSHOT2, LEFT2 and CUT, but you can have any names of your choice.

Your completed wire-frame model (with all text added) should resemble Fig. 3.3 and should be saved as **C:\R13MODEL\WORK3D**.

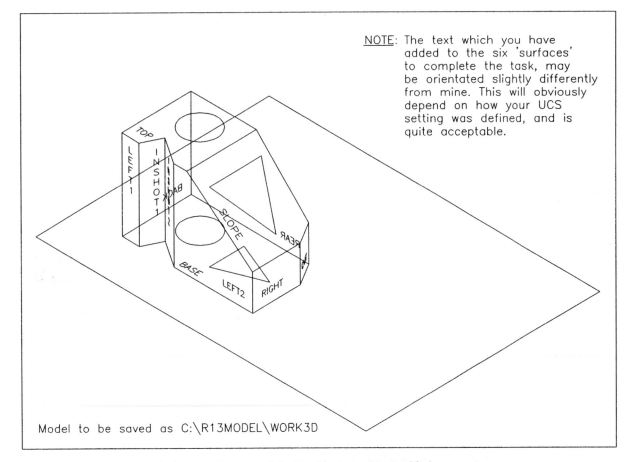

Figure 3.3 The completed wire-frame model with objects and text added.

Notes

1 The user should realize by now that the UCS is an important command with 3D modelling. As stated earlier we have only used the command line entry method of activating this command.

2 Two of the 'surfaces' are the same but have been given two UCS settings and had two items of text added. These are LEFT1 and LEFT2. I thought that this was more correct than just the one LEFT surface, as there is the 'V-notch' cut-out which divides the left-side into two.

3 I have continually used the word 'surface' when referring to the model. All wire-frame models are 'hollow' and do not have surfaces as we know that word, but I hope you understand what is meant at this level when we talk of the top surface, slope surface etc.

C) The UCS options

The UCS command has 14 options available for selection, and we will now discuss each of these with simple exercises using the completed WORK3D drawing, which should still be on the screen.

World: this options returns the WCS irrespective of the current UCS position. It is a very useful option and is the default UCS setting. At the command line enter **UCS** <R> then **W** <R>

Origin: used to set a new UCS origin point. The user specifies the new origin point:
  a) by coordinate input
  b) by referencing existing entities. When used, the icon is positioned at the selected point (if Icon Origin is ON) and assumes the same orientation as the previous icon. We have used this option in our exercise.

ZAxis: defines the UCS position relative to the Z-axis. The user specifies:
  a) the origin point
  b) a point on the Z-axis.
  At the command line enter **UCS** <R> then **ZA** <R>
  prompt   Origin point and **pick INTersection of pt1**
  prompt   Point on positive portion of Z-axis
  respond   **pick INTersection of pt2**
  The UCS icon will be aligned as point 1 with:
  a) the *X*-axis pointing to point 5
  b) the *Y*-axis pointing to point 6
  c) the *Z*-axis pointing to point 2
  Return to WCS with UCS-W.

3point: allows the user to define the UCS position be specifying three points:
  a) the actual origin point
  b) a point on the positive *X*-axis
  c) a point on the positive *Y*-axis
  We used this option to set UCS SLOPE. It is a very easy option to use and understand, and is probably my preferred way of setting a new UCS position.

OBject: aligns the icon to an entity, e.g. a circle, line etc. At the command line enter **UCS** <R> then **OB** <R>
  prompt   Select object to align UCS
  respond   pick any point on top circle

The UCS icon will be aligned:
a) with the origin at the circle centre
b) with the positive *X*-axis towards the circle circumference at the selected point.
The icon can be aligned to lines, polylines, text items, dimensions and blocks.
Try the option with some other entities then return to WCS – easy by now?

View: aligns the UCS so that the *XY*-plane is always perpendicular to the view plane.
a) enter **UCS** <R> then **V** <R>
b) note the icon orientation – similar to 2D?
c) return to WCS.
I find this a useful option as it allows me to add 2D text to a 3D drawing – try it for yourself.

X,Y,Z: allows the UCS to be rotated about the entered axis by an amount specified by the user. We used this option in our exercise.

Prev: restores the previously 'set' UCS position. The option can be used to return the last 10 UCS positions. Try it by entering **UCS** <R> then **P** <R>.

Restore: entering **UCS** <R> then **R** <R> will allow the user to enter a previously saved named UCS. This option was used in the exercise.

Save: again this option was used in the exercise. It is an option which should be used every time a new UCS position has been defined. The user enter **UCS** <R> then **S** <R>
A saved UCS can have up to 31 characters in its name.

Del: Entering **UCS** <R> then **D** <R> prompts with the UCS name(s) to be deleted. The default is none.

?: the option which lists all previously saved UCS's.
At the command line enter **UCS** <R> then **?** <R> and:
prompt UCS name(s) to list<*>
enter * <R>
prompt Text window screen with:
Saved coordinate systems:
BACK
Origin = ... , *X*-axis = ...
Y-axis = ... , *Z*-axis = ...
BASE
Origin = ... , *X*-axis = ...
Y-axis = ... , *Z*-axis = ...
etc.

This is a complete list of all 11 UCS settings. Note the *X*, *Y* and *Z* coordinate values – many are made from the number 0 and 1. More on this in a later chapter.

D) UCS command from the menu bar and toolbar

Until now we have only used the command line entry method of activating the UCS. The only reason for this was that I think it is easier for the new user to understand what option is being used. The UCS command can however be activated by:

a) menu bar selection with **View**
>**Set UCS**

The user is prompted with a cascade menu of the UCS options.

b) icon selection from:
  i) the Standard toolbar – 14 flyout icon options
  ii) the UCS toolbar – 14 icon options. The toolbar is activated from the menu bar with **Tools**
>**Toolbars**
>**UCS**

Note: while the toolbar selection displays 14 icons for selection, two of these are different from the command line options. The options **Delete** (Del) and **List** (?) are omitted, while two new options are available – **Preset UCS** and **Named UCS**.

Figure 3.4 displays the UCS toolbar.

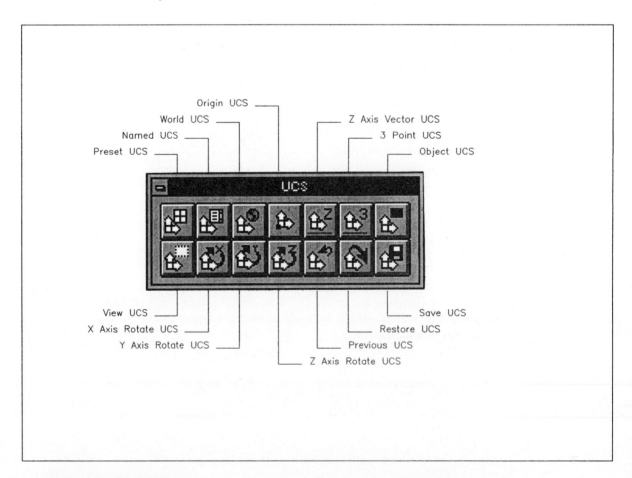

Figure 3.4 The UCS toolbar.

E) The named UCS dialogue box

Restoring a previously saved UCS position is relatively simple, as it involves:

a) entering UCS then R at the keyboard
b) selecting the Restore UCS icon from the toolbar

The problem with both these methods is that the user has to know the name of the UCS which has to be restored. If you have set and saved several UCS positions, you may not remember all of the names. The list (?) option of the UCS command would allow you to see all of the saved names, but the problem is easily overcome by using the UCS dialogue box which lists all the saved settings.

a) at the command line enter **UCS** <R> then **R** <R>
   prompt   ?/Name of UCS to restore
   enter   **BASE** <R>
b) From the menu bar select **View–Named UCS ...**
   prompt   UCS Control dialogue box
   with   1 list of all saved UCS's
          2 BASE current
   respond   1 pick LEFT1 – turns blue
          2 pick Current – gives Fig. 3.5
          3 pick OK
c) The icon is set to the LEFT1 setting.

The UCS Control dialogue box is very useful as it allows:
1 the user to see all the saved UCS names
2 setting of the current UCS, i.e. restore and is easier than the command line method?
3 the delete and list options
4 the ability to rename a UCS by:
   a) pick BASE
   b) alter BASE name in Rename box to ABCD
   c) pick Rename box
   d) BASE name replaced by ABCD in list
   e) pick Cancel.
5 The UCS Orientation dialogue box can be activated from the Named UCS icon from the toolbar.

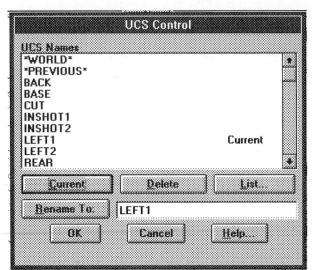

Figure 3.5 The UCS Control dialogue box.

F) The UCS Preset dialogue box

AutoCAD has several preset UCS settings which can be restored from a dialogue box, so:

1 Restore UCS BASE.

2 From the menu bar select **View–Preset UCS ...**
prompt UCS Orientation dialogue box – similar to Fig. 3.6
respond 1 pick FRONT icon
          2 pick OK

3 The UCS icon will be aligned for a 'front' setting.

4 Try the following sequence:
a) select the Preset UCS icon from the toolbar
b) pick Previous then OK – back to BASE
c) select the Preset UCS icon
d) pick RIGHT then OK and icon displayed as a 'right' setting.
e) Select Preset UCS icon-Previous-OK – back to BASE?
f) UCS Preset icon-BACK-OK and icon displayed for the back.
g) Reset to Previous to give the BASE start point.

5 The UCS Orientation dialogue box is useful, but should be used with caution. The FRONT, RIGHT, etc. selection are dependent on the current UCS setting.

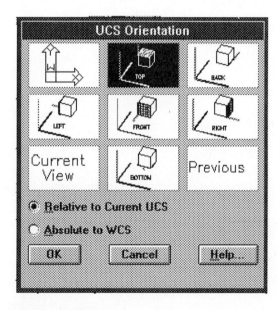

Figure 3.6 The UCS Orientation dialogue box.

G) Plan

Plan is a command which results in a view perpendicular to the XY-plane of the current UCS setting. It is a useful command. We will demonstrate its use with the following sequence:

1 Ensure UCS has been restored to BASE and refer to Fig. 3.7.

2 At the command line enter **PLAN** <R>
prompt <Current UCS>/Ucs/World
enter   **<RETURN>** –, i.e. accept the Current default UCS

3 A plan view of the wire-frame model is displayed – fig. (a). This view is at right angles to the BASE UCS setting, and is really a 'top' view in orthogonal terms.

4 Enter **U** <R> to undo the plan command.

5 Restore UCS LEFT1.

6 From the menu bar select **View–3D Viewpoint Presets–Plan View–Current**

7 The model is displayed as a plan view relative to the LEFT1 UCS setting. It is in fact a view perpendicular to the left side of the model – fig. (b).

8 At the command line enter **PLAN** <R>
   prompt   <Current UCS> ...
   enter   **U** <R> – the ucs option
   prompt   ?/Name of UCS
   enter   **SLOPE** <R>

9 The model is displayed as a view perpendicular to the slope surface as fig(c).

10 From the menu bar select **View–3D Viewpoint Presets–Plan View–Named**
   prompt   ?/Name of UCS
   enter   **CUT** <R>

11 View is now at right angles to the cut surface – fig. (d).

12 Finally restore UCS BASE then from the menu bar select **View–3D Viewpoint Presets–SE Isometric** and we are back to the start of the exercise?

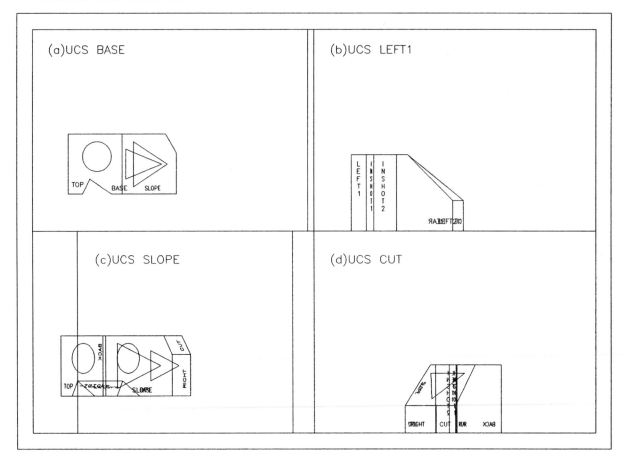

Figure 3.7   PLAN command on wire-frame model.

H) UCSFOLLOW

UCSFOLLOW is a system variable which controls the view display of a model when the UCS position is altered. The variable can only have a value of 0 (default) or 1 and:

0: no effect on the view with UCS changes
1: automatically generates a plan view when the UCS is altered.

As an exercise try the following:

1 UCS back to BASE?

2 At the command line enter **UCSFOLLOW** <R>
prompt   New value for UCSFOLLOW<0>
enter   **1** <R>

3 Nothing has happened?

4 Restore UCS BASE – plan view as fig. (a)

5 Restore UCS LEFT1 – plan view as fig. (b)

6 Restore UCS SLOPE – fig. (c)

7 Restore UCS CUT to give fig. (d).

8 Select the Previous UCS icon from the toolbar three times to return to the UCS BASE plan view.

9 Set the UCSFOLLOW variable back to 0, then select the SE Isometric 3D Preset Viewpoint from the menu bar.

10 This returns our model to the start point as is the last exercise in this chapter.

Summary

1 Wire-frame models are created from coordinate input and by referencing existing entities.
2 The UCS is an important factor in 3D modelling.
3 It is recommended that a UCS setting be made and saved for every 'surface' on a wire-frame model.
4 The UCS command has 14 different options. The option used to set a new UCS position is dependent on the new orientation position.
5 The UCS toolbar offers fast option selection.
6 When setting new UCS positions, ensure that **Icon Origin** is ON.
7 The Named UCS dialogue box is recommended for restoring saved UCS positions.
8 The UCS Orientation dialogue box has preset UCS settings, but should be used with care.
9 PLAN is a command which gives a view perpendicular to the current UCS *XY*-plane.
10 UCSFOLLOW is a variable which can be set to give automatic plan views.

Activity

Creating wire-frame models is important as it allows the user to:

a) use 3D coordinate entry
b) set and save UCS positions.

I have included two wire-frame models which have to be created. The suggested approach is to:

1 Open the **STDA3** standard sheet

2 Make layer MODEL current.

3 Start the model at some convenient point, e.g. 50,50,0 and draw the base – relative coordinate entry is recommended

4 Add the other surfaces one at a time

5 When complete, set and save a UCS for every surface on the model

6 When the model is complete, save as C:\R13MODEL\TUT-2, etc.

The two tutorials are:

Tutorial 2: a mill-guide block. This is a relatively simple model to create, consisting only of horizontal and vertical surfaces.

Tutorial 3: a special slip-block. This model has two sloped surfaces, but is still fairly easy to construct.

Notes:
a) ***Do not attempt to dimension these models***.
b) Remember to set and save UCS positions.
c) Add text items to each surface as shown in the tutorials
d) Do not worry about the two tutorials being displayed on a single page – you will soon be able to do this yourself.

Chapter **4**

Modifying 3D models

All the modify commands which can be used with 2D draughting are available for 3D models, but the results may not be as expected due to the UCS position. We will use our created wire-frame model to investigate the COPY, ARRAY and STRETCH commands.

COPY command

1. Open the **WORK3D** wire-frame model from your directory and refer to Fig. 4.1. Display the Modify and Object Snap toolbars.
2. Ensure that UCS BASE is current.

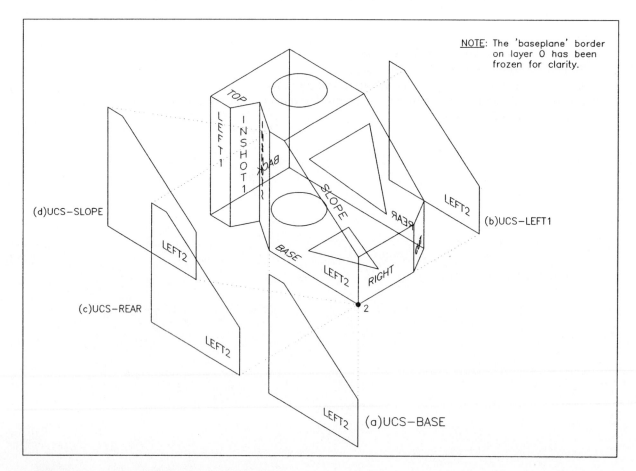

Figure 4.1 WORK3D after the COPY command.

Modifying 3D models **35**

3 Select the COPY icon from the Modify toolbar and:

prompt   Select objects
respond **pick the 5 red lines and green LEFT2 text item** as indicated then right-click
prompt   <Base point ...
respond **pick INTersection of pt2**
prompt   Second point ...
enter   **@0,0,–120** <R> – fig. (a).

4 Restore UCS LEFT1.

5 Select the COPY icon and:
prompt   Select objects
enter   **P** <R> – the previous selection set option which should pick the 6 items as before.
prompt   Base point ... and pick INTersection of pt2
prompt   Second point ... and enter @0,0,–120 <R> – fig. (b).

6 Restore UCS REAR and repeat the COPY command with:
a) the same six entities
b) the same base point, i.e. INT pt2
c) the same second point, i.e. @0,0,–120 – fig. (c).

7 With the UCS restored to SLOPE, repeat the COPY command with the same selection and entries as before – fig. (d).

8 Save if required.

9 The COPY command thus depends on the position of the UCS.

ARRAY command

1 Open **WORK3D** with the UCS BASE current and refer to Fig. 4.2 overleaf.

2 Select the RECTANGULAR ARRAY icon from the Modify toolbar and:
prompt   Select objects
respond **pick the RIGHT text item** then right-click
prompt   Number of rows and enter **5** <R>
prompt   Number of columns and enter **3** <R>
prompt   Row distance and enter **50** <R>
prompt   Column distance and enter **30** <R>

3 The text item is array in a 5 × 3 rectangular matrix as fig. (a).

4 Select the POLAR array icon from the Modify toolbar and:
prompt   Select objects
respond **pick the TOP text item** then right-click
prompt   Center point ... and enter **0,0** <R>
prompt   Number of items and enter **8** <R>
prompt   Angle to fill ... and enter **360** <R>
prompt   Rotate objects ... and enter **Y** <R>

5 The TOP text item is arrayed in a circular pattern – fig. (b).

6 Restore UCS RIGHT.

7 Rectangular array the **SAME** RIGHT text item as before for:
a) 5 rows and 3 columns
b) 50 row distance and 30 column distance
c) result is fig. (c).

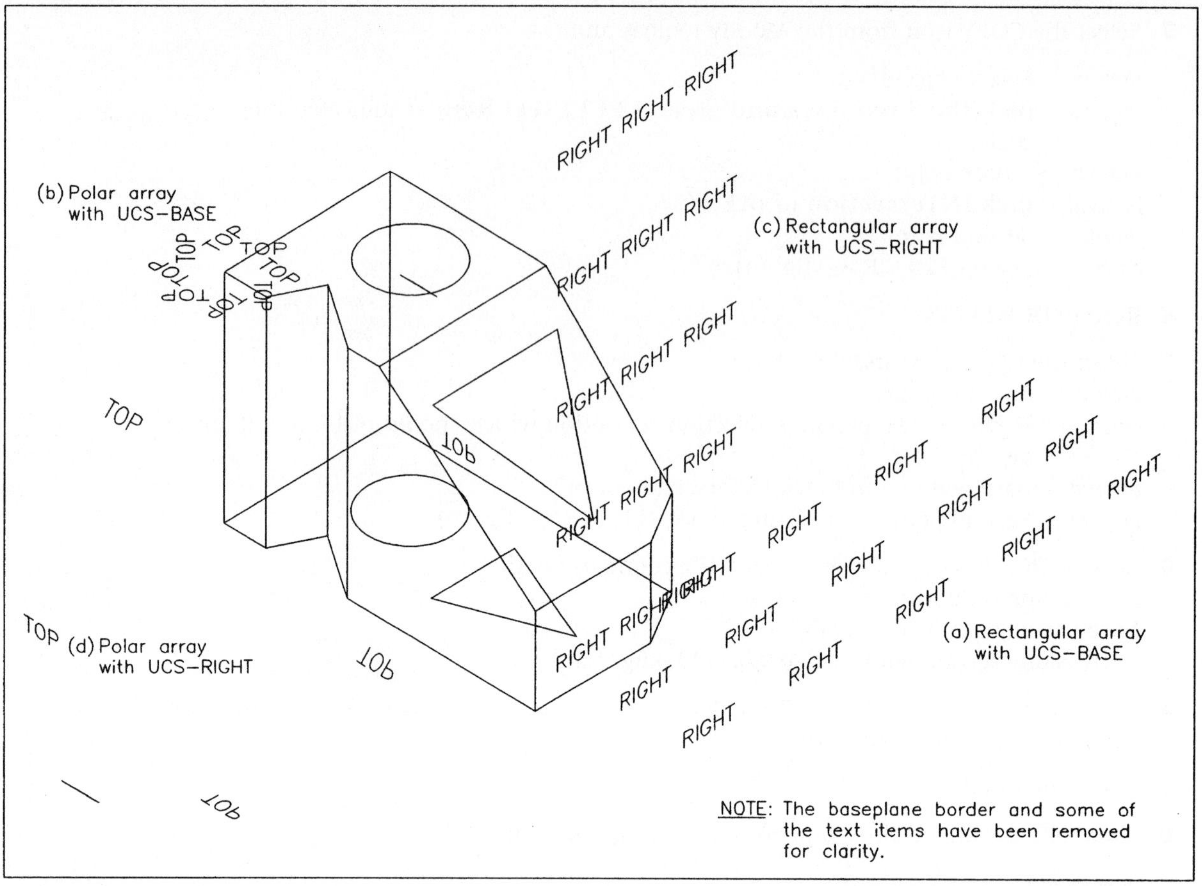

36 *Modelling with AutoCAD*

Figure 4.2 WORK3D after the ARRAY commands.

8 Polar array the same TOP text item as step 4 using:
   a) 0,0 as the array centre point
   b) 8 items, 360 fill angle, with rotation
   c) fig. (d) is the result.

9 Save?

Other commands

Most of the other modify commands can be used with 3D models, e.g. offset, trim, extend, etc. The only requirement is that the UCS is set to the 'plane' for the modification. Certain commands will prompt with the following warning:

View is not plan to UCS
Command results may not be obvious.

Summary

The AutoCAD modify commands with wire-frame models have to be used with care. All commands are available to the user, but the result is dependent on the UCS position. Certain commands give prompts that the result may not be as expected.

Chapter **5**

Dimensioning 3D models

There are no special 3D dimension commands in Release 13. Dimensioning is a 2D concept, the user adding the dimensions to the *XY*-plane of the current UCS setting. This means that the 'orientation' of the complete dimension entity will vary, depending on the UCS position. AutoCAD has automatic dimensioning, and the user should know that the LINEAR dimension will give a horizontal or vertical dimension depending on where the dimension line is located in relation to the entity being dimensioned.

We will demonstrate 3D model dimensioning with two worked examples. The first will be to add dimensions to our wire-frame model WORK3D and the second example will introduce the user to AutoCAD's 'stored' 3D objects.

Dimension example 1

1 Open **WORK3D** from your directory. The wire-frame model will be displayed with green text from the previous chapter and there should be several saved UCS's.

2 Freeze layer TEXT and display the Dimensioning toolbar.

3 Use the Dimension Style dialogue box to create a new dimension style with the following information:
Name: **3DDIMS** from ISO-25

Geometry	Dimension Line	Spacing: 12
	Extension Line	Extension: 3
		Origin Offset: 3
	Arrowheads	Closed Filled
		Size: 6
	Center	None
Format	User Defined: ON	
	Force Line Inside: OFF	
	Fit: Text and Arrows	
	Horizontal justification: Centered	
	Text:	Inside Horizontal: OFF
		Outside Horizontal: ON
	Vertical Justification: Above	
Annotation text	Style:	STD3D – set with standard sheet
	Height:	6
	Gap:	2
	Primary units:	0.00 precision

38 *Modelling with AutoCAD*

4 Refer to Fig. 5.1(a).
   The entities which we will dimension are (a) line 12, (b) line 23, (c) line 14, (d) line 56 and the circle on the top surface. For reference I have added the letters a, b, c and d to the dimension value for reasons which will be become apparent (I hope).

5 Make the **DIMS** layer current and ensure UCS **BASE** is current.

6 *a*) Select the LINEAR dimension icon from the toolbar and:
   prompt   First extension line origin
   respond **INTersection and pick pt1**
   prompt   Second extension line origin
   respond **INTersection and pick pt2**
   prompt   Dimension line location
   respond **pick to suit**
   b) Repeat the LINEAR dimension for line 23
   c) Select the DIAMETER dimension icon from the toolbar and:
   prompt   Select arc or circle
   respond **pick the circle on the TOP surface**
   prompt   Dimension Line Location
   respond **pick to suit** – interesting result?

7 Restore UCS LEFT1, fig. (b), and dimension:
   a) Linear lines 12, 14
   b) Aligned line 56
   c) Try and dimension the circle on the TOP surface

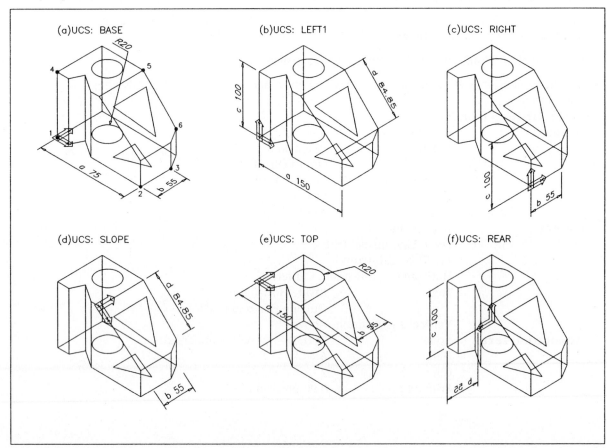

Figure 5.1 Dimensioning WORK3D with various UCS settings.

8 With the UCS restored to RIGHT, dimension:
  a) Linear line 23
  b) Linear line 14 – funny result?
  c) fig. (c)

9 Restore UCS SLOPE and add dimensions to:
  a) Linear 56
  b) Linear 23 – interesting?
  c) fig. (d)

10 With UCS restored to TOP, fig. (e), dimension:
  a) Linear lines 12 and 23
  b) Diameter the top circle.

11 Finally restore UCS REAR and dimension:
  a) Linear lines 14 and 23 – fig. (f)
  b) Note the dimension text orientation.

12 Note: this exercise should demonstrate that:-
  a) adding dimensions in 3D is very dependent on the UCS
  b) there are no special 3D dimension commands
  c) while all line entities can be dimensioned with the LINEAR option, the actual position and orientation of the dimension line is again dependent on the UCS.
  d) the 'correct' 3D dimensions are added to entities on the *XY*-plane of the required UCS.

13 Task: a) erase all dimensions
  b) using layer DIMS and the created dimension style 3DDIMS, refer to Fig. 5.2 and add the dimensions given. You will have to set the UCS correctly.
  c) save as **WORK3D** for future work.

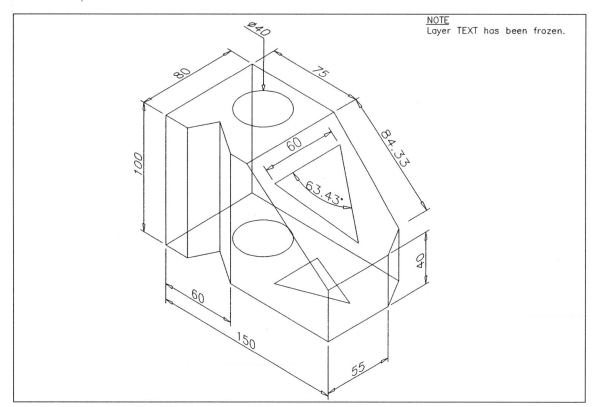

Figure 5.2 Required dimensions for WORK3D.

40 *Modelling with AutoCAD*

Task

Create dimension style 3DDIMS on your STD3D standard sheet. This will save some considerable time when new 3D models have to be dimensioned.

Dimension example 2

In this example we will add dimension to some of Release 13's stored 3D objects, which is quite interesting in itself.

1 Open your **C:\R13MODEL\STDA3** standard sheet with the black border displayed in a 3D configuration.

2 Create the same dimension style 3DDIMS outlined in step 3 of Exercise 1 – you may have already created the 3DDIMS style on your STD3D standard sheet?

3 Select the Save icon from the toolbar. This will automatically update your 3D standard sheet with this dimension style included and current.

4 Ensure layer MODEL is current, and that the WCS is current.

5 Refer to Fig. 5.3

6 From the menu bar select **Draw**
   Surfaces
   3D Objects ...

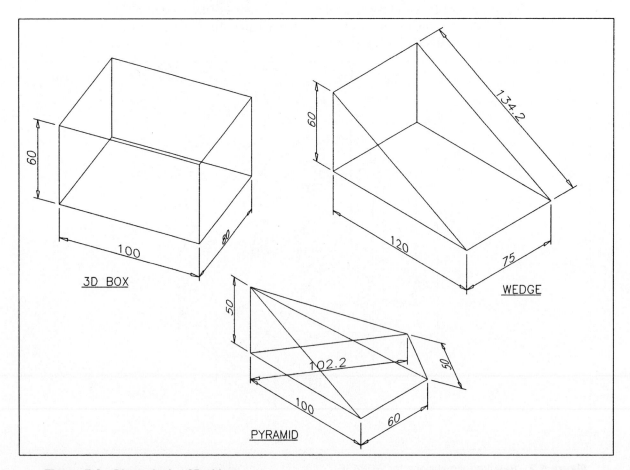

Figure 5.3 Dimensioning 3D objects.

Dimensioning 3D models **41**

prompt	3D Objects dialogue box
with	9 selections (Box-Mesh)
respond	1 pick 3D Box name from list
	a) line turns blue
	b) box icon turns black
	2 pick OK
prompt	Initializing ... 3D Objects loaded
then	Corner of box
enter	**50,50,0** <R>
prompt	Length and enter **100** <R>
prompt	Cube/<Width> and enter **80** <R>
prompt	a yellow rectangle outline is drawn
then	Height and enter **60** <R>
prompt	Rotation about Z-axis and enter **20** <R>

7 A red box object is displayed.

8 Select from the menu bar **Draw**
                            Surfaces
                            3D Objects ...

prompt	3D Objects dialogue box
respond	**pick the Wedge icon then OK**
prompt	Corner of wedge and enter **150,200,0** <R>
prompt	Length and enter **120** <R>
prompt	Width and enter **75** <R>
prompt	Height and enter **60** <R>
prompt	Rotation about Z-axis and enter **0**.

9 Using the menu bar selection **Tools–Toolbars**, activate the **Surfaces** toolbar and position it to suit on the drawing screen.

10 Select the PYRAMID icon from the surfaces toolbar and:

prompt	First base point and enter **250,25,0** <R>
prompt	Second base point and enter **@100,0,0** <R>
prompt	Third base point and enter **@60<90,0** <R>
prompt	Tetrahedron/<Fourth base point>
enter	**@50<150,0** <R>
prompt	Ridge/Top/<Apex point>
enter	**250,25,50** <R> – the apex point.

11 Task You now have to add the dimensions (all LINEAR) given in Fig. 5.3. This requires you to set several UCS positions and I will not give you any help, other than to say that I used 2 for the box, 3 for the wedge and 4 for the pyramid.

12 Save your work when the dimensions are added.

42 *Modelling with AutoCAD*

Summary

1 There are no special 3D dimensioning commands.

2 Dimensioning a 3D model involves adding the dimensions to the *XY*-plane of the required UCS setting.

3 If the UCS is not positioned correctly, dimensions can have the wrong orientation and dimension line location in relation to the entity being dimensioned.

Activity

I have included two old and two new activities for you to dimension.

1 Add the dimensions to Tutorial 2 and Tutorial 3 for the wire-frame models created in chapter 3. This will entail creating the new dimension style 3DDIMS for each model – or try and create your own 3D modelling dimension style.

2 *Tutorial 4*: a simple shaped block. Draw the shaped block to the sizes given, then set and save the four UCS's. The UCS names are listed as suggestions only. Dimension the model using your own dimension style, or create the 3DDIMS style used in the previous exercises. When the model is complete, save as C:\R13MODEL\BLOCK as it will be used in future chapters.

3 *Tutorial 5*: a pyramid. Draw the pyramid as a wire-frame model with a 200 unit square base and a vertical height of 150. Two of the sides are vertical and two are sloped. Set and save the five UCS positions given then add the dimensions. Also add blue circles (layer OBJECTS) to three of the surfaces using the following information:
a) on the base, centre at the square 'centre' with a radius of 40.
b) on the left vertical face, centre at 60,50 relative to the UCS LEFTVERT origin with a radius of 30.
c) on the slope1 face, centre at 80,60 relative to UCS SLOPE1 origin. The radius is to be 40.
Save the completed model as C:\R13MODEL\PYRAMID for recall in later chapters.

Chapter **6**

Hatching in 3D

There are no special 3D hatch commands. Hatching (like dimensioning) is a 2D concept, the hatch pattern being added to the *XY*-plane of the current UCS setting. As usual we will demonstrate hatching by worked examples.

Hatch example 1

1 Open the 3D standard sheet STDA3 with Draw, Modify and Object Snap toolbars. Refer to Fig. 6.1

2 With layer MODEL current, draw the following four planes using the LINE command:
Face	1234	Face	1564	Face	3467	Face	6789
From	30,30,0	From	30,30,0	From	30,130,100	From	30,130,100
To	@100,0,0	To	@0,0,100	To	@100,0,0	To	@0,100,0
To	@0,100,0	To	@0,100,0	To	@0,0,-100	To	@100,0,0
To	@-100,0,0	To	@0,0,-100	To	<RETURN>	To	@0,-100,0
To	close	To	<RETURN>			To	<RETURN>

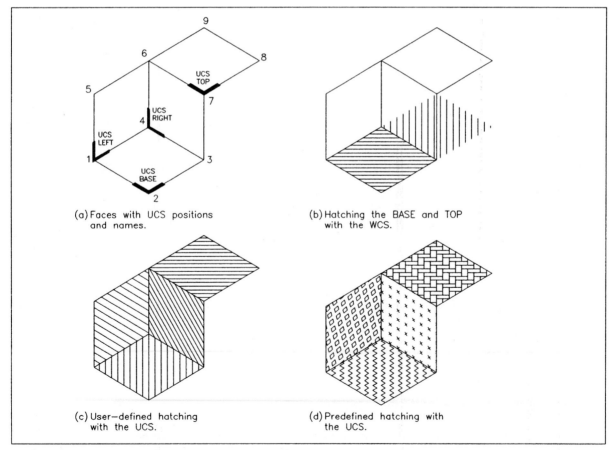

(a) Faces with UCS positions and names.

(b) Hatching the BASE and TOP with the WCS.

(c) User-defined hatching with the UCS.

(d) Predefined hatching with the UCS.

Figure 6.1 Three-dimensional hatch example 1.

44 Modelling with AutoCAD

3 Erase the black border.

4 Set and save the four UCS positions using the names from fig. (a) and make layer HATCH (cyan) current.

5 Ensure WCS is current and select the HATCH icon from the Draw toolbar. Using the Boundary Hatch dialogue box:
  a) set to User-defined Pattern type
  b) set Angle to 45
  c) set Spacing to 8
  d) using the **Pick Points** option:
   i)   pick a point within the 1234 face then right-click
   ii) **Preview–Continue–Apply**.

6 Repeat the HATCH icon selection, then:
  a) use the Pick Points option to pick a point within the 6789 face and:
   prompt   Boundary Definition Error message as Fig. 6.2
   respond pick OK then right-click
  b) use the Select Objects option to pick the four lines of the 6789 face then right-click and:
   i)   set Angle to –45
   ii) Preview–Continue–Apply

7 The result of the two hatch operations are displayed in fig. (b). Face 1234 has the correct hatching, but face 6789 has none, the hatching having been added to the WCS plane.

8 Use the HATCH command again and try and add hatching to the two vertical faces. Possible?

9 Restore UCS BASE and with the HATCH icon:

  a) select the Pick Points option

  b) pick a point within the 1234 face

  c) set Angle to 45 and Spacing to 8

  d) Preview–Continue–Apply.

10 With UCS restored to TOP, hatch the top surface using the pick point option with the angle set to –45.

11 Now use the Select Objects option of the HATCH command to add hatching to the two vertical faces, remembering to set the UCS to LEFT and RIGHT as appropriate.

12 The result of these hatch operations is shown in fig. (c).

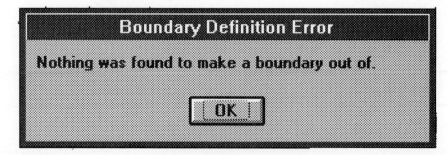

Figure 6.2 The Boundary Definition Error dialogue box.

13 Erase all the hatching then use the predefined hatch patterns to add hatching to the four faces using:

UCS	Pattern	Scale	Angle
BASE	ZIGZAG	30	0
LEFT	SQUARE	25	15
RIGHT	CROSS	20	45
TOP	AR-HBONE	1	0

14 These hatch pattern additions are shown in fig. (d).

15 This completes exercise 1 which does not have to be saved.

Hatch example 2

1 Open the working drawing **WORK3D** and freeze layers TEXT and DIMS.

2 Make layer HATCH current and refer to Fig. 6.3.

3 Restore UCS TOP, select the HATCH icon and:
   a) select the Predefined Pattern STARS
   b) set the Scale to 25 and Angle to 0
   c) use the Pick Points option and pick a point within the top surface boundary
   d) Preview–Continue–Apply.

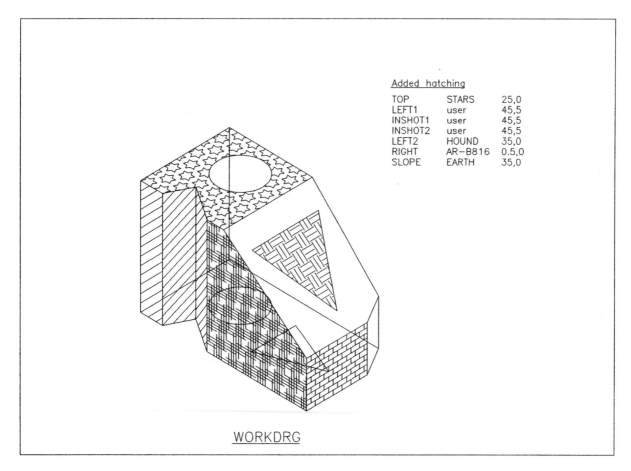

Figure 6.3 Hatching added to WORK3D.

46 *Modelling with AutoCAD*

4 Restore UCS LEFT1 and with the HATCH command:
 a) select User-defined pattern
 b) set the Angle to 45 and Spacing to 5
 c) use the Select Objects option and pick the four lines of face LEFT1
 d) Preview–Continue–Apply.

5 Repeat the hatch command on using the following information:

UCS	Pattern	Scale	Angle	Spacing	Type
INSHOT1	User		45	5	Select Objects
INSHOT2	User		45	5	Select Objects
LEFT2	HOUND	35	0		Select Objects
SLOPE	EARTH	35	0		Pick Points – triangle
RIGHT	AR-B816	0.5	0		Select Objects

6 From the menu bar select **Tools**
 Hide

7 The HIDE command does not affect the hatched wire-frame model as it is still 'hollow'. Note however the circle on the top surface has no back edge line through it – interesting?

8 Save your hatched model as C:\R13MODEL\WORK3D.

Note

In our two worked examples, we used the one layer (HATCH) for all the hatching which was added. This is ***not recommended***. Hatching should be added on individual layers, i.e. there should be a layer made for every UCS position. This will then allow individual hatch layers to be frozen, which may be necessary to allow the selection of other hatch boundary entities.

Summary

1 Hatching is a 2D concept.
2 Hatching a 3D wire-frame model involves setting the UCS to the face which is to be hatched.
3 Both the Pick Points and Select Objects options are both available, although the Pick Points option has to be used with caution.
4 When hatching a wire-frame model it is **STRONGLY RECOMMENDED** that a hatch layer be made for every surface which is to be hatched. I use the same name as the saved UCS positions for my hatch layers, e.g. LEFTHATCH, BASEHATCH, etc.
5 When using the R13 stored hatch patterns, take extra care with the scale factor. Hatching can cause even the largest disc to become 'full' if too small a scale factor is used. Always start with a large scale factor and use the Preview hatch option to reduce the value to give a reasonable hatch effect.
6 A saved hatched 3D wire-frame model can use a lot of memory.

Activity

I have included two hatch activities, both of which involve already created wire-frame models. In both activities, you have to add hatching to several faces of the model using individual hatch layers.

Tutorial 6: the wire-frame BLOCK model.

1 Open the BLOCK model saved during the dimensions chapter.

2 Freeze layer DIMS

3 Create the following new layers (colour CYAN):

LEFTHATCH, RIGHTHATCH, SLOPE1HATCH, SLOPE2HATCH

4 Making each new hatch layer current, restore the appropriate UCS setting and add the following hatch patterns:

Layer	UCS	Pattern
LEFTHATCH	LEFT	HEX, scale 35, angle 0
RIGHTHATCH	RIGHT	TRIANG, scale 30, angle 0
SLOPE1HATCH	SLOPE1	BOX, scale 30, angle 45
SLOPE2HATCH	SLOPE2	HONEY, scale 40, angle 0

5 When completed save your hatched model as it will be used in the viewport and viewpoint chapters.

Tutorial 7: the PYRAMID wire-frame model.

1 Open the PYRAMID model and freeze layer DIMS.

2 Make the following new layers (colour CYAN):

BASEHATCH, LEFTVERTHATCH, RIGHTVERTHATCH, SLOPE1HATCH, SLOPE2HATCH

3 With the correct UCS and layer current, add the following hatching to the five surfaces of the model:

Layer	UCS	Pattern
BASEHATCH	BASE	AR-PARQ1, scale 3, angle 0
LEFTVERTHATCH	LEFTVERT	AR-B816, scale 1, angle 0
RIGHTVERTHATCH	RIGHTVERT	AR-B816, scale 1, angle 0
SLOPE1HATCH	SLOPE1	AR-B816, scale 1, angle 0
SLOPE2HATCH	SLOPE2	AR-B816, scale 1, angle −

Note:

a) No hatching has to be added to any of the circles on the faces.

b) You will probably have to freeze each hatch layer after the hatching has been added to allow entities to be selected for the next hatch surface.

c) When all surfaces are hatched, save your model.

Chapter **7**

Viewports

The graphics screen can be divided into a number of separate viewing areas called **viewports**, and each viewport can show any part of a drawing. Viewports are **interactive**, i.e. what is drawn in one viewport is automatically drawn in the others and the user can switch between viewports when constructing a model. Viewport layouts – called configurations – can be saved for recall, thus allowing different configurations of the same model to be stored in memory. Viewports are essential when working in 3D as they allow different views of the model to be displayed on the screen at the one time. When used with the VIEWPOINT command (next chapter), the user has a very powerful draughting tool.

Release 13 has two types of viewports:

1 Tiled or fixed
2 Untiled or floating.

The type of viewport which is displayed at any one time is controlled by the **TILEMODE** variable, and:

a) Tilemode 1: tiled viewports – cannot be moved (default)
b) Tilemode 0: untiled viewports – can be moved

Figure 7.1 displays the two types of viewport.

In this chapter we will only investigate TILED viewports and leave the floating viewports to a later chapter when we will investigate the concept of model and paper space.

The viewport command can be activated from the menu bar or by direct entry from the command line, and in this chapter we will consider both methods.

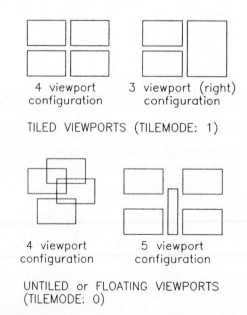

Figure 7.1 Tiled/untiled viewports.

Example 1: WORK3D

This example will investigate viewports using an already created model – our 3D working drawing WORK3D.

1 Open the WORK3D wire-frame model which should be displayed with hatching, text and possibly dimensions. Delete all dimensions and freeze layers TEXT and HATCH.

2 Make layer MODEL current and restore the WCS.

3 At the command line enter **TILEMODE** <R> and:
prompt   New value for TILEMODE – probably 1 default
enter   **1** <R> – tiled option

4 From the menu bar select **View**
                      Tiled Viewports
                      Layout ...
prompt   Tiled Viewport Layout dialogue box
respond 1 pick **vport–3r** from list and:
               a) it turns blue
               b) icon display turns black – Fig. 7.2
          2 pick OK

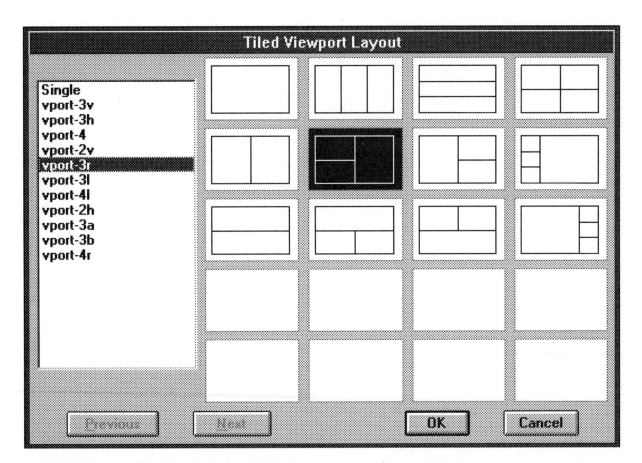

Figure 7.2 The Tiled Viewport Layout dialogue box.

50 *Modelling with AutoCAD*

5 The drawing screen will be returned and be divided into three separate areas, one large to the right (the 3r) and two smaller to the left as Fig. 7.3(a). The three viewports all display the same view of WORK3D but at different 'sizes'.

Note: you may have to reposition the toolbars?

6 Moving the pointing device about the screen will give:
   a) the large viewport displays the cursor cross-hairs and is the ACTIVE viewport, i.e. current.
   b) the other two viewports display an arrow and these viewports are NON-ACTIVE.

7 An individual viewport is made active by:
   a) moving the pointing device into the viewport area
   b) left-clicking.
   Try this for yourself a few times.

8 From the menu bar select **View**
   Tiled Viewports
   Save
   prompt   ?/Name for new viewport configuration
   enter   **CONF1** <R>

9 Make the upper left viewport active and from the menu bar select **View–Tiled Viewports–2 Viewports** and:
   prompt   Horizontal/<Vertical>
   enter   **V** <R> – the vertical (default) option

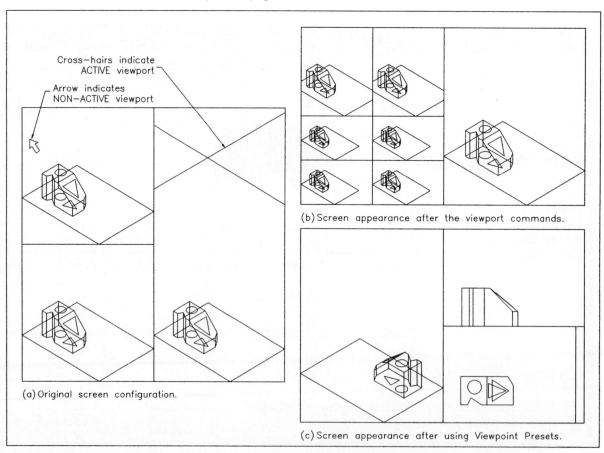

Figure 7.3 Viewport example 1.

10 The top left viewport will be further divided into two equal vertical viewports, each displayed the wire frame model.

11 With the lower left viewport active, select from the menu bar **View–Tiled Viewports–4 Viewports**, and the lower left viewport will be further 'split' into four equal viewports, all displaying the model.

12 At this stage your screen should resemble Fig. 7.3(b).

13 At the command line enter **VPORTS** <R> and:
   prompt   Save/Restore/Delete ...
   enter   **S** <R> – the save option
   prompt   ?/Name for new viewport configuration
   enter   **CONF2** <R>

14 Now make the large right viewport active and from the menu bar select **View–Tiled Viewports–1 Viewport**. Your screen should display the original WORK3D drawing – I had to ZOOM-ALL?

   Save this configuration as **CONF3**, i.e. View–Tiled Viewports–Save.

15 At the command line enter **VPORTS** <R>
   prompt   Save/Restore ...
   enter   **3** <R> – the 3 viewport option
   prompt   Horiz/Vert/Above ...
   enter   **L** <R> – the left option

16 Restore UCS BASE and with the upper right viewport active select from the menu bar
   View
   3D Viewpoint Presets
   Front

17 With the lower right viewport active, select from the menu **View**
   3D Viewpoint Presets
Plan View
Current

18 Make the large left viewport active then **View**
   3D Viewpoint Presets
NW Isometric

19 Your drawing screen should be the same as Fig. 7.3(c) and we are displaying a top, front and rear isometric of our wire-frame model. This has been achieved without any additional drawing on our part – we only had the original 3D view at the start of the exercise.

   Note:
   a) the size of the model in the viewports may be different, and this will be investigated later in this chapter.

   b) the viewpoint command will be investigated fully in the next chapter.

20 Now save this new configuration as CONF4.

21 Finally make the large left viewport active and then select from the menu bar **View–Tiled Viewports–1 Viewport** to return to the original screen layout (zoom all?)

52 *Modelling with AutoCAD*

22 From the menu bar select **View**
Tiled Viewports
Restore

prompt ?/Name of viewport configuration to restore
enter **?** <R> – the list option
prompt Viewport configuration(s) to list
enter ***** <R> – the all option
prompt Text screen with:
Current configuration
id#2
corners: 0.0000, 0.0000, 1.0000, 1.0000
Configuration CONF1:
0.5000, 0.0000, 1.0000, 1.0000
etc.
then ?/Name of ...
enter **CONF1** <R>

23 The first saved viewport configuration will be displayed.

24 At the command line enter **VPORTS** <R>
prompt Save/Restore ...
enter **R** <R>
prompt ?/Name of ...
enter **CONF4** <R>

25 Restore the other configurations then save if you want, but **not** as WORK3D. This completes the first exercise.

Example 2

The first viewport exercise used an already created 3D model to investigate the viewport command and configurations. This exercise will create a wire-frame model using a 4 viewport configuration with views preset. We will 'see' the model being created in all four viewports.

1 Open your STD3D standard sheet with MODEL layer current. Refer to Fig. 7.4 which displays stages in the construction of the wire frame model.

2 Check tilemode is set to 1.

3 Set a new UCS origin position by selecting from the menu bar **View**
Set UCS
Origin

prompt Origin point<0,0,0>
enter **50,50,0** <R>

4 Save this UCS as BASE.

5 The UCS icon should move to this new origin point. If it did not, then check from the menu bar **Options–UCS** and:
a) Icon ON, i.e. tick
b) Icon origin ON, i.e. tick

6 Set a 4 viewport configuration by selecting **View**
Tiled Viewports
4 Viewports

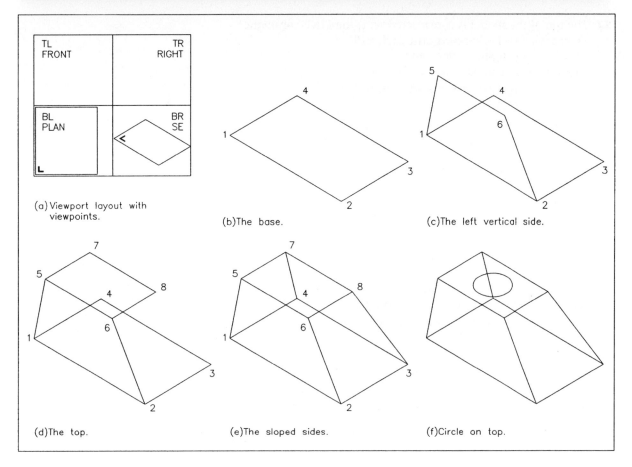

Figure 7.4 Construction of viewport example 2.

7 Making the appropriate viewport active in turn, use **View–3D Viewpoint Presets** to set the following views:
   a) top left (TL) viewport – Front
   b) top right (TL) viewport – Right
   c) bottom left (BL) viewport – Plan view – Current
   d) bottom right (BR) viewport – SE Isometric

8 The viewport and viewpoint commands are listed in fig. (a).

9 With the bottom right (BR) viewport active, construct the base of the model with the LINE command:
   From point   **0,0,0** <R>   – pt1
   To point   **@200,0,0** <R>   – pt2
   To point   **@0,120,0** <R>   – pt3
   To point   **@200<180,0** <R>   – pt4
   To point   **c** <R>   – fig. (b).

10 Using the LINE command, construct the left vertical side:
   From point   **INTersection and pick pt1**
   To point   **@20,0,100** <R>   – pt5
   To point   **@120,0,0** <R>   – pt6
   To point   **INTersection and pick pt2**
   To point   right-click   – fig. (c)

11 The top of the model is constructed with the LINE command:
From point **INTersection and pick pt5**
To point **@0,80,0** <R> – pt7
To point **@120,0,0** <R> – pt8
To point **INTersection and pick pt6**
To point right click – fig. (d).

12 Now add the 'sloped sides' by joining points 4–7 and 3–8 with the LINE command to give fig. (e).

13 Make layer OBJECTS (blue) current and draw a circle:
a) centre at 80,40,100
b) radius 25.
The blue circle is 'added' to the top surface in all four viewports – fig. (f).

14 From the menu bar select **Draw**
     Surface
     3D Objects ...
     3D Box then OK
prompt Corner of box and enter **80,30,0** <R>
prompt Length and enter **50** <R>
prompt Cube/<Width> and enter **40** <R>
prompt Height and enter **30** <R>
prompt Rotation ... and enter **15** <R>

15 A blue cuboid is displayed on the base surface inside the wire-frame model and the complete four viewport configuration display should be the same as Fig. 7.5.

16 At this stage you may want to save your work as **TEST3D**, but do not exit the drawing.

REDRAW and REGEN

REDRAW is a command which clears the screen of unwanted blips and pixels which have been 'left behind' from certain commands, e.g. ERASE. The command is generally activated by picking the REDRAW icon from the Standard toolbar. When working in multi-screen mode, the REDRAW command is only effective in the current viewport, but it is possible to redraw all viewports with the one command REDRAWALL. This command can be activated from the menu bar or from the redraw flyout menu:

a) Menu bar: View
     Redraw View – active viewport
     Redraw All – all viewports
b) Icon: Redraw View   Redraw All

REGEN is a similar command to regen, but gives a 'better definition' of circular objects. The command must be entered from the keyboard and can be viewport or global, i.e.

REGEN: active viewport

REGENALL: all viewports.

Centring a model

The second exercise created a wire-frame model in a four viewport configuration as Fig. 7.5, but the model is displayed in each viewport at different 'sizes'. There are several ways of 'centring' models within viewports and we will now consider one of these methods – zoom centre.

1 Ensure UCS BASE is current.

2 With the lower right viewport active, erase the four black border lines.

3 In each viewport select **View–Zoom–All**.

4 Make the lower left viewport active.

5 The model is basically a shaped cuboid of overall sizes 200 × 120 × 100 and its 'centre point' is thus at 100,60,50. These figures are relative to the UCS BASE position.

6 From the menu bar select **View**
                                     Zoom
                                     Center
prompt   Center point
enter   **100,60,50** <R>
prompt   Magnification or Height<297>
enter   **200** <R>

7 The plan view is displayed at a larger scale and is centred in the viewport.

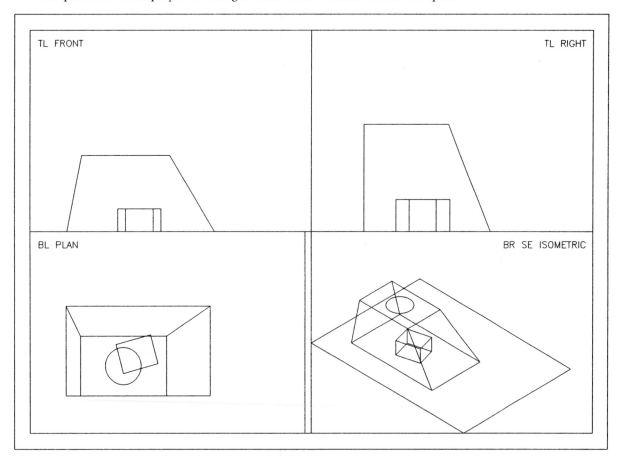

Figure 7.5 Four viewport configuration for example 2.

8 Use the Zoom–Center option in the other three viewports with:
  a) 100,60,50 as the center point
  b) 200 as the magnification in the top left and top right viewports
  c) 225 as the magnification in the bottom right (3D) viewport.

9 The result should be Fig. 7.6, i.e. the model is centred in each viewport.

10 This completes the exercise and the multi-screen display can now be saved as **TEST3D**, overwriting the previous saved model.

Summary

1 Viewports can be tiled (fixed) or untiled (floating).
2 The viewport mode is determined by the TILEMODE variable and:
TILEMODE 1 – tiled viewports (default)
TILEMODE 0 – untiled viewports
3 Using viewports gives multi-screen configurations.
4 Tiled viewports can have between 1 and 4 screen displays.
5 The viewpoint command can be used to set top, front, right, etc. views within individual viewports.
6 Multi-screen layouts are extremely useful with 3D modelling.

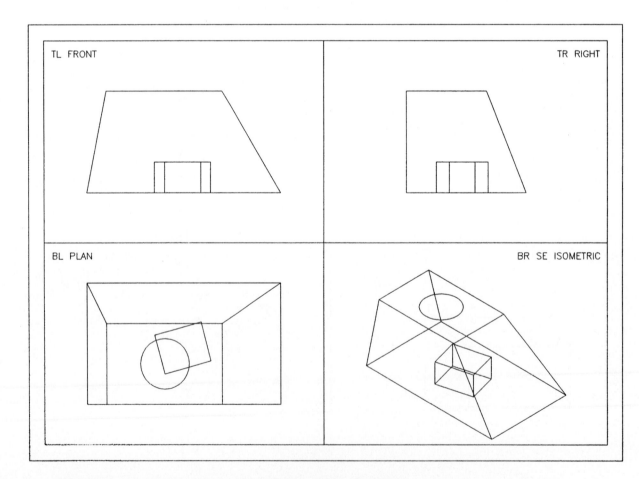

Figure 7.6 Viewport example 2 after the zoom centre commands.

Activity

There are three activities for you to attempt the models having already been created. These activities will be used in the next two chapters, so make sure they are saved.

Tutorial 8: the Special Slip Block – Tutorial 3

1 Open the drawing and erase any dimensions.

2 Freeze layer TEXT.

3 Set a UCS as shown and save as BASE.

4 Display a three viewport (Right) configuration.

5 Save viewport configuration as **VPEX1**

Tutorial 9: the hatched cuboid – Tutorial 6.

1 Open the drawing and erase all dimensions.

2 Restore UCS BASE.

3 Set the viewport configuration to three (Above).

4 Save as **VPEX2**

Tutorial 10: the hatched pyramid – Tutorial 7.

1 Open the drawing and erase dimensions.

2 Set UCS to BASE

3 Display a four viewport configuration.

4 Save as **VPEX3**.

Note

The appearance of the models in the viewport configuration drawings is not ideal as:
1. they are the same
2. they are not centred in individual viewports.

The next two chapters will investigate these two topics.

Chapter **8**

Viewpoint

Viewpoint is the command that determines how the user 'looks at' a model. We have used the command in previous chapters without any discussion about it and in this chapter we will investigate the command in detail using some of our previously saved models. When the viewpoint command is combined with viewports, the user has a very powerful draughting tool – multiple viewport displays.

The viewpoint command has four different options, these being:

1 the ROTATE option

2 the TRIPOD option – bull's eye and target

3 the VECTOR option

4 the 3D VIEWPOINT PRESETS.

The actual command can be activated from the menu bar or by direct entry from the keyboard.

Viewpoint rotate option

This option requires two angle values:

a) the angle in the *XY*-plane from the *X*-axis – the view direction
b) the angle from the *XY*-plane – the inclination (tilt).

1 Open the WORK3D wire-frame model.

2 Make layer MODEL current and thaw layer TEXT. Freeze layers HATCH and DIMS.

3 Make UCS BASE current and set the 3D Viewpoint Preset to SE Isometric – it probably is set to this.

4 Refer to Fig. 8.1 (section A) which displays several views of the model with the angle from the *XY*-plane set to 0 (prompt 2).

5 At the command line enter **VPOINT** <R>
    prompt   ***Switching to the WCS***
    then   Rotate/<View point><1.00, -1.00, 1.00>
    enter   **R** <R> – the rotate option
    prompt   Enter angle in *XY*-plane from *X*-axis<315>
    enter   **40** <R>
    prompt   Enter angle from *XY*-plane<35>
    enter   **0** <R>
    prompt   ***Returning to the UCS***
    then   Regenerating drawing.

6 The model is displayed looking onto the right/rear sides from a horizontal 'stand-point' – fig. (a1).

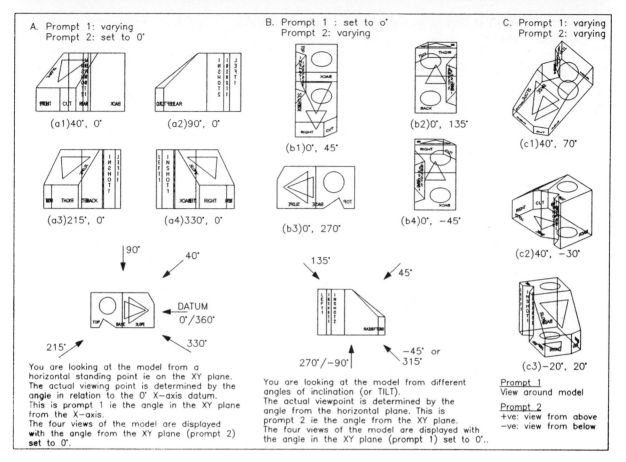

Figure 8.1 Viewpoint – the ROTATE option.

7 Repeat the keyboard **VPOINT R** command and enter the following angle values at the prompts:

prompt 1	prompt 2	fig.
90	0	a2
215	0	a3
330	0	a4

8 Restore the original SE Isometric viewpoint.

9 Refer to Fig. 8.1 (section B) which displays several viewpoints of the model with the angle in the *XY*-plane from the *X*-axis set to 0, i.e. prompt 1.

10 Using the keyboard **VPOINT R** command, enter the following angles at the prompts:

prompt 1	prompt 2	fig.
0	45	b1
0	135	b2
0	270 or –90	b3
0	–45 or 315	b4

11 Again restore the original SE Isometric viewpoint.

12 Refer to Fig. 8.1 (section C) which displays three views of the model with varying angle values for both prompts.

60 *Modelling with AutoCAD*

13 Use the **VPOINT R** keyboard entry with the following angles:

prompt 1	*prompt 2*	*fig.*
40	70	c1
40	−30	c2
−20	20	c3

14 Restore the SE Isometric viewpoint.

15 Restore some other saved UCS positions, e.g. LEFT1, SLOPE and repeat the VPOINT R command using some of the above entries. The final display of the model is not affected by the UCS position.
Think about the prompt ***Switching to the WCS***

16 Explanation of option
   a) Prompt 1: Angle in the *XY*-plane from the *X*-axis.
   This is your actual standpoint **on the horizontal plane** as you look at the model, i.e. it is your view direction. If this angle is 0° you are looking at the model from the datum position and looking to the right side of the model. If the angle is 270° you are looking onto the front of the model. The value of this angle can be between 0° and 360°. It can also be positive or negative. Remember that 270° is the same as −90°.
   b) Prompt 2: Angle from the *XY*-plane.
   This is your actual 'head inclination' looking at the model, i.e. it is your angle of tilt. A 0° value means that you are looking at the model horizontally and a 90° value is looking vertically down. The angle of tilt can vary between 0 and 360 and be positive or negative and:

   +ve tilt: looking down on the model

   −ve tilt: looking up at the model.

Viewpoint Rotate from menu bar

The viewpoint rotate option can be activated from the menu bar, allowing the user to set the angle prompts via a dialogue box.

1 WORK3D model still displayed?

2 Ensure UCS BASE is current and the viewpoint has been restored to SE Isometric

3 From the menu bar select **View**
   3D Viewpoint
   Rotate ...
   prompt   Viewpoint Presets dialogue box as Fig. 8.2.

4 This dialogue box contains the following:
   a) viewing angle is **absolute to WCS**
   b) angle from *X*-axis is 315 and is displayed in the left-hand 'clock' diagram
   c) angle from *XY*-plane is 35.3 and is displayed in the right-hand 'arc quadrant' diagram.

5 The dialogue box allows:
   a) viewing angles to be set   *i*) **absolute to WCS**
   ii) **relative to UCS**
   b) angles to be set by selecting the circle/arc position
   c) angles to be set by altering the values at the **From:** line
   d) plan views to be set.

6 Respond to the dialogue box with:
   a) change the *X*-axis angle from 315 to 150
   b) change the *XY*-plane angle from 35.3 to 10
   c) pick OK.

7 The model is displayed at the new viewpoint.

8 Make UCS SLOPE current.

9 Select from the menu bar **View–3D Viewpoint–Rotate** and using the dialogue box:
   a) change to **Relative to UCS** – black dot
   b) pick OK.

10 The model viewpoint is altered.

11 Make UCS BASE current and use the Viewpoint Presets dialogue box to:
   a) restore Absolute to WCS
   b) set *X*-axis angle to 315
   c) set *XY*-angle to 35.3
   d) pick OK.

12 The model is now displayed at its original viewpoint.

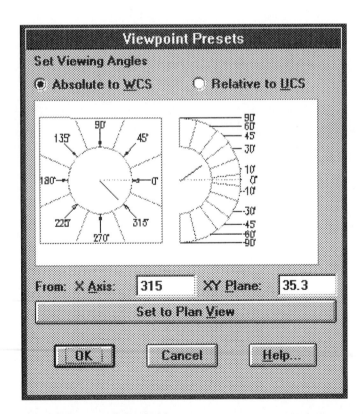

Figure 8.2 The Viewpoint Presets dialogue box.

62 Modelling with AutoCAD

Viewpoint tripod option

This option allows the user the ability to set 'infinite viewpoints'. While the option may appear to be complex, but is actually very easy to use.

1 Ensure that WORK3D is displayed at the SE Isometric viewpoint and that the UCS BASE is current.

2 Refer to Fig. 8.3 which displays several views of the model using this option.

3 From the menu bar select **View**
   3D Viewpoint
   Tripod

   prompt   XYZ axes tripod and the bulls-eye and target cross. The axes and cross move as the pointing device is moved.

   respond move the target (+) into the circle quadrant indicated in fig. (a) and left-click

4 The model is displayed at this viewpoint.

5 Using the tripod option, position the target in the different quadrants indicated in Fig. 8.3, i.e.
   (a)–(d): within inner circle
   (e)–(h): within outer circle.

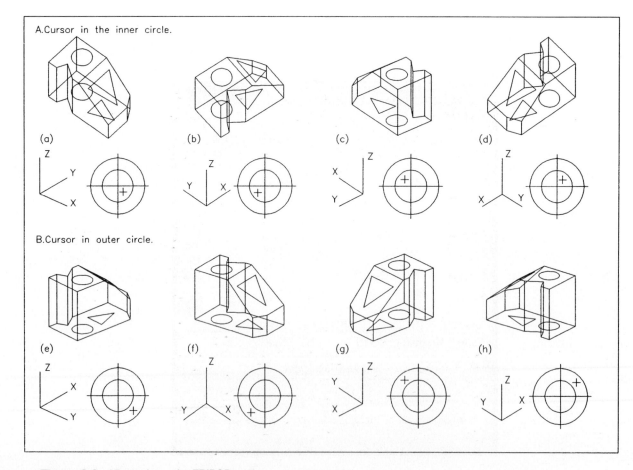

Figure 8.3 Viewpoint – the TRIPOD option.

6 When you are confident at using the tripod, try the following:

a) position the target at different points on the axes and note the viewpoint display

b) position the target on the circle circumferences and observe the viewpoints obtained

c) restore some other saved UCS positions and use the tripod option. You should find that the viewpoint display is not dependant on the UCS position.

7 Explanation of option

a) The 'bulls-eye' is in reality a representation of a glass globe and the model is located at the centre of this globe. The *XY*-plane is positioned at the equator. The north pole of the globe is at the circle centres and the two concentric circles represent the surface of the world, stretched out onto a flat plane with:

circle centre: the north pole

inner circle: the equator

outer circle: the south pole.

b) As the target is moved about the circles, you are moving around the surface of the globe and:

Cursor position	View result
in inner circle	above equator, looking down on model
in outer circle	below equator, looking up at model
on inner circle	looking horizontally at model
below horizontal	viewing from the front
above horizontal	viewing from the rear

Viewpoint Vector option

For this option we will use another model – only as a change from WORK3D, so:

1 Open **TEST3D**, the wire-frame model created during the exercise on viewports. The screen should display a four viewport configuration with the model 'centred' in each viewport?

2 Restore the WCS.

3 Refer to Fig. 8.4 and make the top left (TL) viewport current.

4 From the menu bar select **View**
                     3D Viewpoint
                     Vector
prompt   Rotate/<Viewpoint><0.00,–1.00,0.00>
enter   **0,0,1** <R>

5 A top view of the model results – fig. (a) and 'fills the viewport'.

6 Make the top right (TR) viewport active then from the menu bar select **View–3D Viewpoint–Vector** and:
prompt   Rotate/<View ...
enter   **0,–1,0** <R>

7 A view on the front of the model is displayed – fig. (b) and fills the viewport.

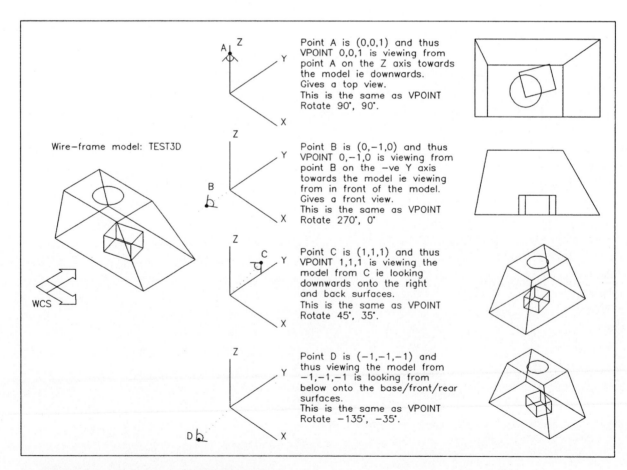

Figure 8.4   Viewpoint – the VECTOR option.

Viewpoint **65**

8 With the bottom left viewport active, enter **VPOINT** <R> at the command line and:
prompt   Rotate ...
enter   **1,1,1** <R>

9 The model is displayed in 3D as fig. (c), and is being viewed from above.

10 Finally in the bottom right viewport, activate the vector option of the Viewpoint command and enter −1,−1,−1 as the 3D vector to give fig. (d) – viewed from below.

11 The original model viewport configuration display is now altered in each viewport. The model is not 'centred' but we will investigate this in the next chapter so save the viewport layout as **TEST3D_A**.

12 **Explanation of option**
    _a_) The vector option allows the user to enter _x, y, z_ coordinates for a 'stand point' looking at the model, i.e. if you enter 0,0,1 you are 'standing' at the point 0,0,1 looking at the model. This point is on the positive Z-axis, so you are looking down onto the top of the model.
    _b_) The actual numerical value of the vector entered does not matter, e.g. 0,0,1; 0,0,10; 0,0,99.99; 0,0,246 are all the same as far as the viewpoint is concerned. I prefer to use 1's, hence 0,1,0; 1,0,0; −1,−1,−1, etc.

3D viewpoint presets

The 3D viewpoint presets have been used in previous chapters and allow the user:
_a_) four isometric viewpoints, e.g. SE Isometric
_b_) six 'traditional' viewpoints, e.g. top, front
_c_) plane views which can be relative to:
    _i_)   the current UCS
    _ii_) the WCS
    iii) a named UCS.

The option can be activated from:

1 the menu bar

2 The View toolbar which gives the four isometric and six traditional viewpoints, but not the plan view.

It is the user's preference as to which of these two methods is used to activate the viewpoint presets. My preference is to use the menu bar selection.

Viewpoint: absolute to WCS or relative to UCS?

When setting a viewpoint, the user must decide whether the viewpoint is to be:

_a_) absolute to the WCS
_b_) relative to a UCS.

When absolute to the WCS (the default setting), the resulting view of the model is completely independent of the UCS position. If the viewpoint is relative to a UCS, the resultant view will obviously depend on the UCS position and may not be as the user would expect. The Viewpoint Rotate dialogue can be used to alter the Absolute/Relative option. It is strongly recommended that all viewpoints are **absolute to the WCS** and that the relative to the UCS is not used.

66 *Modelling with AutoCAD*

The VIEW command

Different views of a model can be saved as views within a drawing, thereby allowing the user to create a series of useful 'pictures' of the model. These pictures can be recalled at any time. To demonstrate the command:

1 Open WORK3D – the hatched wire-frame model. Erase any dimensions and restore the WCS – it may be current?

2 At the command line enter **VIEW** <R>
  prompt   ?/Delete/Restore ...
  enter   **S** <R> – the save option
  prompt   View name to save
  enter   **ORIGINAL** <R>

3 Using the 3D Viewpoint Presets, set to NW Isometric.

4 At the command line enter **VIEW** <R>
  prompt   ?/Delete ... and enter **S** <R>
  prompt   View name to save and enter **V1** <R>

5 Using the 3D Viewpoint Presets, set a Front view, then use the VIEW command to save the view as V2.

6 At the command line enter **VIEW** <R>
  prompt   ?/Delete/Restore ...
  enter   **?** <R>
  prompt   View(s) to list<*> and right-click
  prompt   Text screen with:
       Saved view   Space
       ORIGINAL   M
       V1   M
       V2   M
  respond F2 to flip back to the drawing screen.

7 From the menu bar select **View**
                    Named Views ...
  prompt   View Control dialogue box
  respond 1 pick ORIGINAL
       2 pick Restore
       3 pick OK.

8 The original view of the model is displayed on the screen.

9 Restore *a*) view V1 from the View Control dialogue box and *b*) view V2 from the VIEW keyboard command.

Summary

1 Viewpoint allows models to be viewed from different points.
2 When viewports and viewpoint arc combined, multiple views of a model are obtained.
3 The viewpoint command has several options – rotate, the tripod, vectors and the presets.
4 The rotate option:
  a) allows models to be viewed relative to two angles:
    i) the angle 'around' the model – the direction
    ii) the angle of inclination or tilt

b) the angle of tilt allows the model to be viewed:
   i) from above if +ve angle value
   ii) from below if −ve angle value.
5 The tripod option allows unlimited viewpoints.
6 The vector option requires:
   a) an *x, y, z* coordinate entry.
   b) an understanding of the right-hand rule.
7 The 3D Viewpoint Presets allow 'set' viewpoints to be easily obtained.
8 Viewpoints can be:
   a) absolute to the WCS – *recommended*
   b) relative to the UCS – *not recommended*.
9 Wire-frame models exhibit **ambiguity** when the viewpoint command is used, i.e. it is difficult to know if you are viewing the model from above or below.

Activity

There are three activities with this chapter, these being the viewport activities from the previous chapter. The activities are very easy and should not take too much of your time. Make sure they are saved for the next chapter.

Tutorial 11

1 Open VPEX1 of the Special Slip Block in a three viewport (right) configuration.

2 Using the Viewpoint–Rotate option, set the following viewpoints:

Viewport	Direction	Tilt	
Top left	270	0	– Front
Bottom left	0	90	– Top
Right	225	35	

3 Set a UCS as indicated in the right viewport. Save as BASE.

4 Save as **VPEX1**.

Tutorial 12

1 Open VPEX2 of the hatched Cuboid which should be displayed as a three viewport (above) configuration.

2 Use Viewpoint with the Vector option to set:

Top:	1, −1, 1	– Isometric
Bottom left:	0, −1, 0	– Front
Bottom right:	−1, 0, 0	– Right.

3 Save as **VPEX2**

Tutorial 13

1 Open VPEX3, the hatched pyramid as four viewports.

2 Using the Viewpoint Presets set to:

Top left:	Front
Top right:	Left
Bottom left:	Top
Bottom right:	SW Isometric

3 Save as **VPEX3**.

Chapter **9**

Centring viewports

When 3D models are displayed in multiple viewport configurations, three 'problems' can initially occur:

1 The model may 'fill the viewport'.

2 The model may be displayed at different scales in the various viewports.

3 The model may not 'line up' between viewports.

These 'problems' are easily overcome by zooming the model about a specified centre point, and this was demonstrated in an earlier worked example with viewports. We will consider two further worked examples to demonstrate the feature.

Example 1: WORK3D

1 Open the WORK3D drawing which should be displayed in 3D with hatching. Freeze the text and dimensions layers and erase the black border.

2 Ensure UCS BASE is current.

3 From the menu bar select **View–Tiled Viewports–4 Viewports**.

4 Using the 3D Viewpoint Presets set the following viewpoints in the specific named viewports:

Top left: Front   Top right: Left

Bottom left: Top Bottom right: SE Isometric.

5 The model has a basic cuboid type shape, the overall size being $150 \times 80 \times 100$. The 'cuboid centre' is thus at the point 75,40,50 **relative to the UCS BASE** position.

6 From the menu bar select **View**
                    Zoom
                    Center

prompt   Center point
enter   **75,40,50** <R>
prompt   Magnification or Height<180?>
enter   **150** <R>

7 Use the Zoom–Centre option in the other three viewports with:
a) 75,40,50 as the centre point
b) 150 as the magnification (200 in the 3D viewport).

8 The result should be as Fig. 9.1.

9 At this stage save the display if required.

10 Restore UCS LEFT1 and zoom-centre in each viewport using the same values as before, i.e.
a) centre point: 75,40,50
b) magnification: 150 or 200.

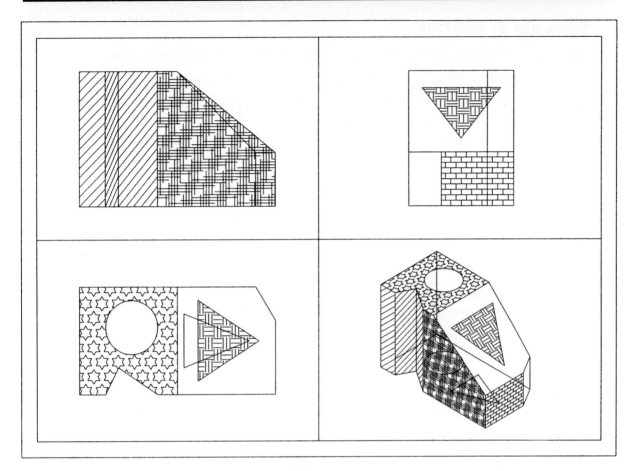

Figure 9.1 Centre viewport example 1 – WORK3D.

11 The result of this zoom–centre command is that the model is not centred in the viewports. The 75,40,50 centre point is relative to the UCS BASE position, and is not suitable for the UCS LEFT1 position. If the UCS is set to LEFT1, the required centre point is 75,50,–40. Try this centre point entry, and the model should be displayed centred in each viewport. Why these coordinates?

12 Restore the WCS and try and centre the viewports.

Example 2: TEST3D

1 Open drawing **TEST3D_A** of the wire-frame model in a four viewport configuration.

2 Ensure UCS BASE is current.

3 The cuboid has basic dimensions of 200 × 120 × 100 so the 'cuboid centre' is 100,60,50 relative to the UCS BASE position.

4 With the top left viewport active, enter **ZOOM** <R> at the command line and:
prompt   All/Center ...
enter   **C** <R> – the center option
prompt   Center point
enter   **100,60,50** <R>
prompt   Magnification or Height<138?>
enter   **180** <R>

5 Use the zoom–center option in the other viewports with:
a) 100,60,50 as the centre point
b) 180 as the magnification in the top right viewport
c) 250 as the magnification in the bottom two viewports.

6 The final result should be as Fig. 9.2.

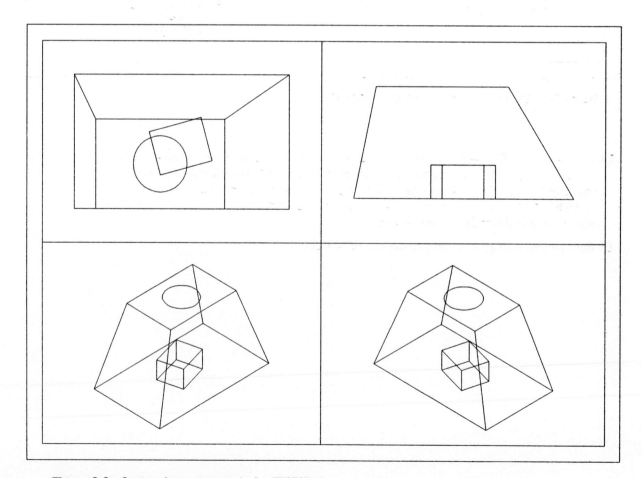

Figure 9.2   Centre viewport example 2 – TEST3D_A.

Summary

1 Models can be centred in viewports using the Zoom–Centre command, the user entering:
   a) a centre point
   b) a magnification.
2 The entered centre point is **relative to the current UCS**.
3 The magnification value entered 'scales' the model relative to the given default, e.g. if the default is <180> then:
   a) a value less than 180 will increase the model size
   b) a value greater than 180 will decrease the size
   c) the scale effect is dependent on the actual value.

Activity

There are three model centre activities these being the saved tutorials from the previous viewpoint chapter.

Tutorial 14

1 Open VPEX1 which displays the Special Slip Block in a three viewport (right) configuration.

2 Ensure UCS BASE is current and zoom centre about the point 90,50,55 at about 150 magnification.

Tutorial 15

Open the three viewport (above) configuration of the hatched cuboid block – VPEX2. This model is basically $150 \times 80 \times 80$.

Tutorial 16

Centre the hatched pyramid VPEX3 – no help with centre point, but watch the 3D viewport. The centre point is not what you might expect due to the model orientation. Think about it!

Chapter **10**

Surface modelling

The best way to describe a surface model is to think of a wire-framed model with 'skins' covering all the wires from which the model is constructed. These surfaces give surface models certain advantages over wire-frame models:

1 the model can be displayed with hidden line removal

2 there is no ambiguity

3 the model can be shaded and rendered.

AutoCAD Release 13 adds **FACETED** surfaces using a polygon mesh technique, but this mesh only approximates curved surfaces. The different types of surface models which are available with R13 are:

- 3D faces
- 3D meshes
- ruled surfaces
- tabulated (extruded) surfaces
- revolved surfaces
- edge surfaces
- polyface surfaces
- 3D objects
- elevation and thickness surfaces.

Each type of surface will be considered in a chapter on its own, with the exception of the elevation/thickness surfaces. This has already been covered in Chapter 1 – did you realize that you were creating surface models at this early stage?

The surface commands can be activated:

1 from the menu bar with **Draw**
         Surfaces

2 from the Surfaces toolbar – Fig. 10.1

3 by direct keyboard entry.

During our investigations all methods will be used.

Surface modelling 73

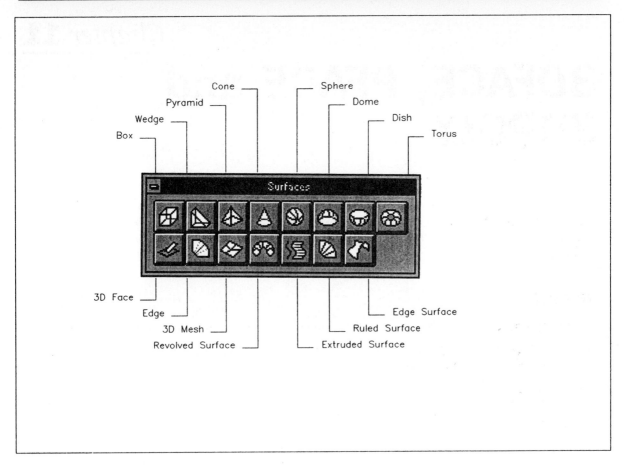

Figure 10.1 The surfaces toolbar.

Chapter 11

3DFACE, PFACE and 3DPOLY

The first two commands can be used to 'convert' a wire-frame model into a surface model, the process being likened to covering a wire-frame model with fabric or skin. A 3DPOLY is a single 3D entity. The commands will be investigated by different worked examples.

3DFACE

To demonstrate how to use this command we will create a new wire-frame model, so:

1 Open the **STD3D** standard sheet to display a black border. Ensure layer MODEL is current and display the Draw, Modify, Object Snap and Surfaces toolbars.

2 Refer to Fig. 11.1 which displays stages in the creation of the surface model.

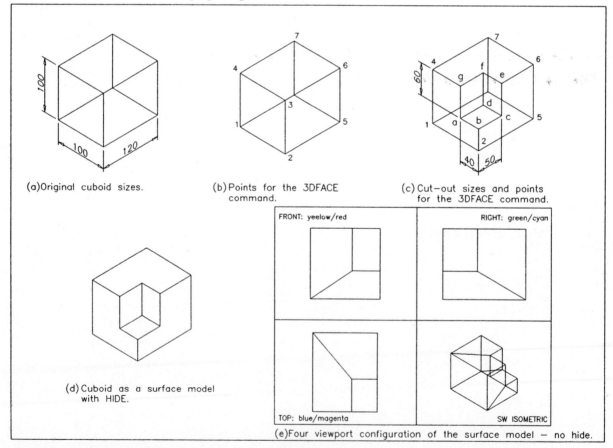

Figure 11.1 3DFACE cuboid surface model example.

3DFACE, PFACE and 3DPOLY

3 Using the LINE icon, create a wire-frame cuboid 100 × 120 × 100 at a suitable point, e.g. 50,50,0 – fig. (a).

4 At the command line enter **HIDE** <R> and nothing happens – the model has no 'sides' to hide.

5 Make a new layer called FACE and make it current.

6 At the command line enter **COLOR** <R>

   prompt   New object color <BYLAYER>
   enter   2 <R> – yellow

7 The colour icon in the Object Properties toolbar is now yellow.

8 Set a running object snap to ENDPOINT and refer to fig. (b).

9 From the menu bar select **Draw**
                            Surfaces
                            3DFACE
   prompt   First point and pick pt1
   prompt   Second point and pick pt2
   prompt   Third point and pick pt3
   prompt   Fourth point and pick pt4
   prompt   a yellow rectangle is displayed on the cuboid face
   prompt   Third point and right-click.

10 At the command line enter **COLOR** <R>
   prompt   New object color<2 (yellow)>
   enter   3 <R> – green

11 Select the 3DFACE icon from the Surfaces toolbar and:
   prompt   First point and pick pt2
   prompt   Second point and pick pt3
   prompt   Third point and pick pt6
   prompt   Fourth point and pick pt5
   prompt   a green rectangle is displayed over the face
   prompt   Third point and right-click.

12 Set the COLOR to 5 (blue) then enter **3DFACE** <R> at the command line:
   prompt   First point and pick pt6
   prompt   Second point and pick pt7
   prompt   Third point and pick pt4
   prompt   Fourth point and pick pt3
   prompt   a blue rectangle outline displayed
   prompt   Third point a right-click.

13 At this stage, we have added three coloured faces to three surfaces of the wire-frame model.

14 From the menu bar select **Tools**
                            Hide

15 The model will be displayed with hidden line removal, i.e. we have converted the original wire-frame model into a surface model.

16 From the menu bar select **Tools**
                            Shade
                            16 Color Filled

76 *Modelling with AutoCAD*

17 The surface model is displayed with coloured surfaces – quite impressive for what has been created?

18 Return to wire-frame representation by entering **REGEN** <R> at the command line.

19 Freeze the MODEL layer, then ERASE the three coloured faces which have been created. Thaw MODEL layer and make it current.

20 Refer to Fig. 11.1(c) and add the 'cut-put' part of the cuboid using the sizes given – draw, erase, trim?

21 With layer FACE current set the colour to yellow, then activate the 3DFACE command and:
prompt　　First point and pick pt1
prompt　　Second point and pick pt2
prompt　　Third point and pick ptb
prompt　　Fourth point and pick pta
prompt　　Third point and right-click.

22 Using the 3DFACE command, add other faces to the cuboid using the following information:
Colour　　Face
yellow　　a-g-4-1
green　　　2-b-c-5 and c-e-6-5
blue　　　6-7-f-e and f-g-4-7
magenta　a-b-c-d
cyan　　　a-d-f-g
red　　　　c-e-f-d

23 Now　　*a*) HIDE the model – fig. (d)
　　　　　　b) SHADE-16 Color Filled – impressive?

24 REGEN and remove the ENDpoint running object snap.

25 Task
　　a) Create a four-viewport configuration and use the 3D Viewpoint Presets to display the model as fig. (e). Hide and shade in each viewport.
　　b) The model is not a 'complete surface model' as three surfaces have not yet been faced. Add the required faces if you want.

26 Save your model if required.

Explanation of the 3DFACE command

The 3DFACE command can be used to face any three or four sided surface. The command can also be used to create 'continuous' faces. To demonstrate this feature of the command, begin a new **2D** drawing, refer to Fig. 11.2 and:

1 Set the snap to 5 or 10 and toggle the grid on.

2 Activate the 3DFACE command and:
prompt　　First point and pick any pt1
prompt　　Second point and pick any pt2
prompt　　Third point and pick any pt3 – 1st pt of next face
prompt　　Fourth point and pick any pt4 – 2nd pt of next face
Face 1-2-3-4 displayed
prompt　　Third point and pick any pt5 – 1st pt of next face
prompt　　Fourth point and pick any pt6 – 2nd pt of next face

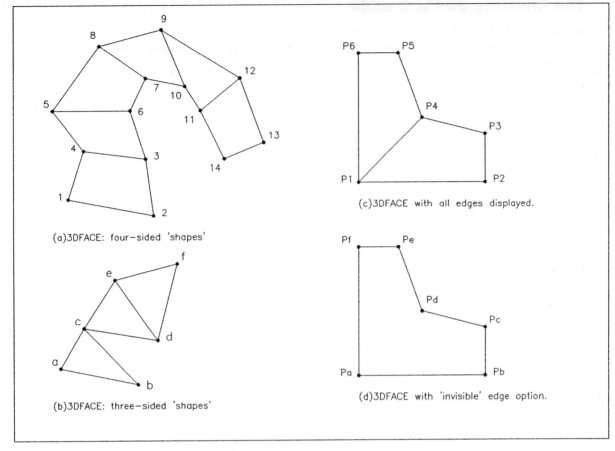

Figure 11.2 Investigating the 3DFACE command.

Face 3-4-5-6 displayed
prompt   Third point and pick any pt7 – 1st pt of next face
prompt   Fourth point and pick any pt8 – 2nd pt of next face
Face 5-6-7-8 displayed
prompt   Third point ...
respond   pick points 9–10; 11–12; 13–14 in respond to the third and fourth point prompts

3 The result should be similar to fig. (a).

4 Activate the 3DFACE command again and:
prompt   First point and pick any pta
prompt   Second point and pick any ptb
prompt   Third point and pick any ptc – 1st pt of next face
prompt   Fourth point and pick the same ptc – 2nd pt of next face
Face a-b-c displayed
prompt   Third point and pick any ptd – 1st pt of next face
prompt   Fourth point and pick any pte – 2nd pt of next face
Face c-d-e displayed
prompt   Third point and pick any ptf
prompt   Fourth point and pick the same ptf
Face d-e-f displayed
prompt   ... and right-click.

5 Result is fig. (b).

78 *Modelling with AutoCAD*

The invisible 3DFACE edge

When the 3DFACE command is used with continuous four-sided shapes, all four sides of the face are displayed. It is possible to create a 3DFACE with an '**invisible edge**'.

1 Refer to Fig. 11.2(c) and draw and L-shaped component similar to that shown – use snap on to help.

2 Copy the component to another part of the screen.

3 Make a new layer called FACE, colour blue and current.

4 Use the 3DFACE command to add faces to:
   a) points P1–P2–P3–P4–right click
   b) points P1–P4–P5–P6–right click

5 The two faces are displayed, but each has an edge between points P1 and P4.

6 Activate the 3DFACE command again and:
   prompt   First point and pick pt Pa
   prompt   Second point and pick pt Pb
   prompt   Third point and pick pt Pc
   prompt   Fourth point
   enter   **i** <R> – for invisible edge
   then   pick pt Pd
   prompt   Third point and right-click.

7 The 3DFACE is displayed without edge Pa–Pd.

8 Repeat the 3DFACE command and:
   prompt   First point
   enter   **i** <R>
   then   pick pt Pa
   prompt   Second point and pick pt Pd
   prompt   Third point and pick pt Pe
   prompt   Fourth point and pick pt Pf
   prompt   Third point and right-click.

9 This face is also displayed without edge Pa–Pd.

10 An invisible edge can be displayed (if required) by selecting from the menu bar **Draw**
                                                                                    Surfaces
                                                                                    Edge

   prompt   Display/<Select Edge>
   respond pick any point on edge Pa-Pb-Pc-Pd
   and   edge between Pa-Pd is displayed
   then   right-click.

11 The invisible edge has been demonstrated in 2D, but can also be used with 3D models.

3DFACE example 2

In this example we will create a 3D wire-frame model and then add faces with invisible edges. The construction of the model is given in Fig. 11.3, so:

1 Open your 3D standard sheet with layer MODEL current.

2 Create a 100 unit cube at a suitable point and then a 60 unit on two of the 'sides' – fig. (a).

3 Make the following new layers:

FACE1 green; FACE2 blue; FACE3 yellow.

4 With layer FACE1 current, refer to fig. (b) and use the 3DFACE command to add faces to 1–2–3–4 and 3–4–5–6.

5 Use the HIDE command – fig. (c), then REGEN.

6 Make layer FACE2 current and with the 3DFACE command:
   prompt First point and enter **i** <R>
   then   pick pt 2
   prompt Second point and pick pt a
   prompt Third point and enter **i** <R>
   then   pick pt d
   prompt Fourth point and pick pt 3
   prompt Third point and right-click

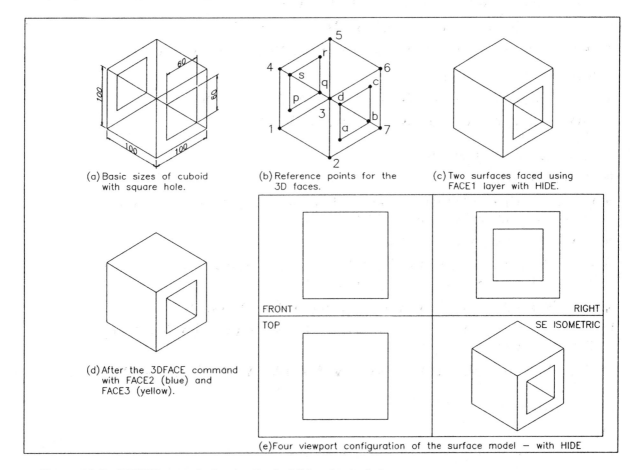

(a) Basic sizes of cuboid with square hole.

(b) Reference points for the 3D faces.

(c) Two surfaces faced using FACE1 layer with HIDE.

(d) After the 3DFACE command with FACE2 (blue) and FACE3 (yellow).

(e) Four viewport configuration of the surface model – with HIDE

Figure 11.3 3DFACE example 2 using the invisible edge technique.

80 *Modelling with AutoCAD*

7 Two blue edges of face 2-a-b-3 are displayed (2–3 and a–d)

8 Using the 3DFACE command with the invisible edge technique used in step 5, add faces to:
a) 3–d–c–6
b) 6–c–b–7
c) 7–b–a–2

9 With layer FACE3 current add faces to:
a) a–b–q–p
b) b–c–r–q

10 At this stage HIDE the model to give fig. (d) then SHADE. REGEN to return the surface model to wire-frame representation.

11 Task
a) Create a 4 viewport configuration as fig. (e).
b) HIDE–SHADE in each viewport then REGENALL.

12 Save if required.

PFACE

A PFACE is a polygon mesh and is similar to a 3DFACE but allows the user to define a number of vertices for the surface (not the 3/4 allowed with the 3DFACE command). The best way to demonstrate the command is by example so:

1 Open your STD3D standard sheet, with layer MODEL (red) current.

2 Make two new layers, DF colour blue and PF colour green and refer Fig. 11.4.

3 Rotate UCS about *X*-axis by 90°.

4 Set an elevation of 0 and a thickness of –200; remeber how? – fig. (a).

5 Using the polygon icon, create a HEPTAGON (7 sides) with:
a) centre at 300,100
b) inscribed in a 50 radius circle.

6 Zoom in on the prism.

7 From the menu bar select **Tools–Hide** and the prisms will appear as fig. (a). As the model is an extrusion there is no 'top or bottom' surface.

8 Make the new layer DF (blue) current.

9 Use the 3DFACE icon to face the top end of the prism then HIDE – fig. (b). I used the command three times with its invisible edge option.

10 Undo the 3DFACE effect with U<R> and make layer PF (green) current.

11 The 'top surface' of the prism has seven edges and we will add a mesh to this surface as a series polygons (triangles).

Note: the command sequence which follows is long, tedious and will seem very repetitive. Persevere!

3DFACE, PFACE and 3DPOLY 81

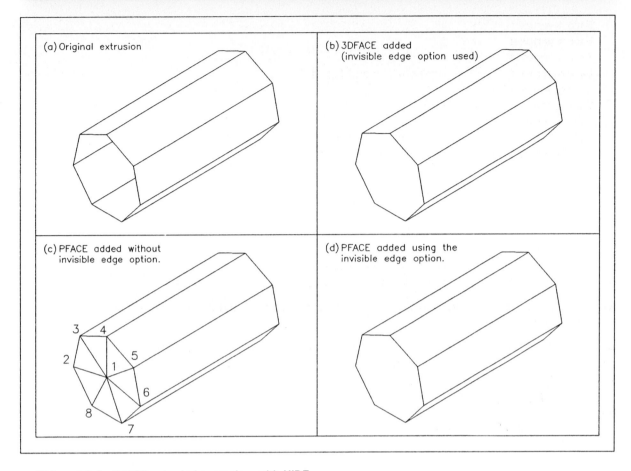

Figure 11.4 PFACE extruded example – with HIDE.

12 At the command line enter **PFACE** and:

prompt	enter or pick
Vertex 1:	300,100 – surface 'centre point' pt1
Vertex 2:	pick INT pt2
Vertex 3:	pick INT pt3
Vertex 4:	pick INT pt4
Vertex 5:	pick INT pt5
Vertex 6:	pick INT pt6
Vertex 7:	pick INT pt7
Vertex 8:	pick INT pt8
Vertex 9:	right-click as no more vertices.
Face 1,vertex 1	1, i.e. vertex 1, the centre point
Face 1,vertex 2	2
Face 1,vertex 3	3
Face 1,vertex 4	right-click, i.e. end of face 1
Face 2,vertex 1	1
Face 2,vertex 2	3
Face 2,vertex 3	4
Face 2,vertex 4	right-click to end face 2
Face 3,vertex 1	1
Face 3,vertex 2	4
Face 3,vertex 3	5
Face 3,vertex 4	right-click to end face 3

Face 4,vertex 1	1
Face 4,vertex 2	5
Face 4,vertex 3	6
Face 4.vertex 4	right-click to end face 4
Face 5,vertex 1	1
Face 5,vertex 2	6
Face 5,vertex 3	7
Face 5,vertex 4	right-click to end face 5
Face 6,vertex 1	1
Face 6,vertex 2	7
Face 6,vertex 3	8
Face 6,vertex 4	right-click to end face 6
Face 7,vertex 1	1
Face 7,vertex 2	8
Face 7,vertex 3	2
Face 7,vertex 4	right-click to end face 7
Face 8,vertex 1	right-click to end command

13 The end surface of the prism will be face with seven green triangular polygon meshes. Hide to give fig. (c).

14 The added PFACE has lines between the sides and the centre, but as with the 3DFACE command, these edges can be made invisible.

15 Erase the added mesh surface.

16 Use the PFACE command with the following:
a) vertex prompts: as step 12
b) face, vertex prompts:

1,1: -1	2,1: -1	3,1: -1	4,1: -1
1,2: 2	2,2: 3	3,2: 4	4,2: 5
1,3: -3	2,3: -4	3,3: -5	4,3: -6
5,1: -1	6,1: -1	7,1: -1	
5,2: 6	6,2: 7	7,2: 8	
5,3: -7	6,3: -8	7,3: -2	

17 The green PFACE will be displayed with invisible edges and will hide as fig. (d).

18 By entering a negative vertex, the edge from that vertex will be invisible, e.g.
Face 1,vertex 1: -1 edge 1–2 is invisible
Face 1,vertex 2: 2 edge 2–3 is visible
Face 1,vertex 3: -3 edge 3–1 is invisible.

19 This completes the PFACE example.

3DPOLY

A 3D polyline is a continuous entity created in 3D space and is similar to a 2D polyline. It does not possess the 2D versatility, i.e. there is no width or arc options with a 3D polyline. A 3D polyline can be edited. To investigate the command:

1 Open your STD3D standard sheet a set a four viewport configuration as given in Fig. 11.5.

2 Create four new layers:
  a) 2DP: red
  b) 3DP_1: blue
  c) 3DP_2: green
  d) SPL: magenta

3 Make the lower left (top) viewport active and layer 2DP current.

4 Draw a 2D polyline:
  From point   50,50
  To point   @0,100
  To point   @10,0
  To point   @−20,50 then right-click.

5 The 2D polyline is drawn on the *XY*-plane – fig. (a).

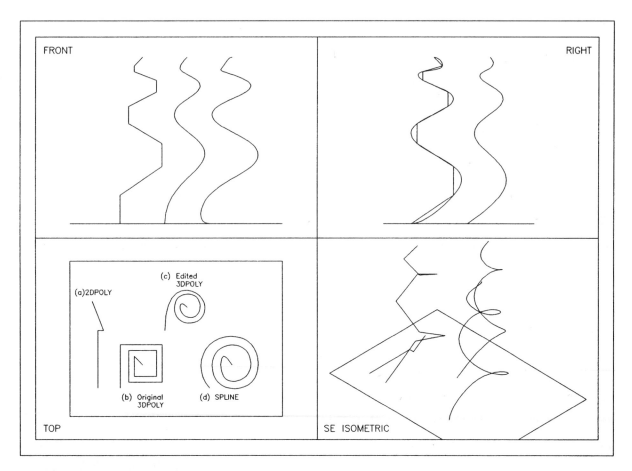

Figure 11.5 3DPOLY/SPLINE example.

84 *Modelling with AutoCAD*

6 Now draw a 2D polyline:
From point 50,50
To point @10,100,100

Message: 2D point or option keyword required, i.e. you cannot use the PLINE command to create 3D polylines.

7 Cancel the PLINE command and make layer 3DP_1 (blue) current.

8 From the menu bar select **Draw**
3D Polyline

prompt	From point and enter 90,50
prompt	Endpoint of line and enter @0,75,50
prompt	Endpoint of line and enter @75,0,50
prompt	Endpoint of line and enter @0,–65,40
prompt	Endpoint of line and enter @–60,0,35
prompt	Endpoint of line and enter @0,55,30
prompt	Endpoint of line and enter @50,0,25
prompt	Endpoint of line and enter @0,–45,20
prompt	Endpoint of line and enter @–40,0,15
prompt	Endpoint of line and enter @0,35,10
prompt	Endpoint of line and enter @15,–15,15
prompt	Endpoint of line and right-click.

9 A blue polyline 'straight spiral' is displayed – fig. (b).

10 Centre each viewport about the point 190,135,150 at about 350 magnification.

11 With layer 3DP_2 (green) current, draw a 3D polyline from the point 170,150 using the same relative coordinate entries as step 8.

12 From the menu bar select **Modify**
Edit Polyline

prompt	Select polyline
respond	pick the green 3D polyline
prompt	Close/Edit vertex ...
enter	**S** <R> – spline curve option
prompt	Close/Edit ...
enter	**X** <R>

13 The green 3D polyline is displayed as a splined curve – fig. (c).

14 Make layer SPL (magenta) current and from the menu bar select **Draw-Spline** and:

prompt	Object/<Enter first point>
enter	**250,50** <R>
prompt	Enter point and enter @0,75,50
prompt	Enter point and enter @75,0,50
prompt	...
enter	the sequence as step 8 then right-click
prompt	Enter start tangent and enter 280,20,0
prompt	Enter end tangent and enter 340,20,290

15 A magenta splined spiral is drawn – fig. (d).

16 Note the shape of the SPLINE curve relative to the splined 3D polyline.

17 This completes the 3D polyline example.

Summary

1 The 3DFACE and PFACE commands allow **surface models** to be created by drawing 'skins' over wire-frame models.

2 The HIDE command allows surface models to be displayed with hidden line removal.

3 REGEN returns a surface model to wire-frame representation.

4 The 3DFACE command can only be used with three/four-sided figures.

5 Continuous faces can be created.

6 The PFACE command is used for multi-sided figures.

7 Faces can be created with **invisible** edges.

8 The **Surface–Edge** command allows invisible edges to be displayed.

9 It is recommended that separate layers are made for each face being 'added' to a model. This allows individual face colours to be set.

10 A 3D polyline is a single entity drawn in 3D space.

11 A 3D polyline can be edited but cannot be drawn with arc segments or variable width.

12 3D polylines are useful for hatching wire-frame models which have difficult surfaces – think about this!

Activity

Tutorial 17 is an interesting activity – a wedge of cheese!

1 Create a wire-frame model of the cheese.

2 3DFACE (or PFACE) the cheese using five new coloured layers.

3 Set a 4 viewport configuration with viewpoints which allow every face of the cheese to be seen. Remember to centre the viewports.

4 Hide and shade in each viewport.

5 Save your work as C:\R13MODEL\CHEESE.

Chapter 12

3DMESH

A 3D polygon mesh consists of a series of 3D faces in an $M \times N$ rectangular matrix. The user enters the x, y, z coordinates of the matrix (M, N) **vertices**. The mesh matrix is defined by

M: columns in the x-direction
N: rows in the y-direction

The values of M and N can be between 2 and 256 and a 3D mesh is a single entity.

Worked example

1 Open the STD3D standard sheet with layer MODEL current.

2 Erase the black border.

3 Select the 3Dmesh icon from the Surfaces toolbar and:
   prompt Mesh M size and enter **5** <R>
   prompt Mesh N size and enter **4** <R>
   prompt Vertex(0,0) and enter **50,50,30** <R>
   prompt Vertex(0,1) and enter **50,120,15** <R>
   prompt Vertex(0,2) and:
   respond Refer to Fig. 12.1 and enter the required coordinate values in response to the vertex prompts

4 When the last vertex coordinate is entered (4,3), the 3D mesh will be displayed as fig. (a).

5 One of the vertex coordinates has been entered wrongly, this being vertex (4,0)

6 Set a four viewport configuration using the viewpoints given in Fig. 12.1(b), and make the lower left (top) viewport active.

7 From the menu bar select **Modify**
                   Edit Polyline
   prompt Select polyline
   respond **pick any point on the mesh**
   prompt EditVertex/Smooth ...
   enter **E** <R> – the edit vertex option
   prompt Vertex(0,0). Next/Previous ...
   and an X is displayed at vertex(0,0) in all viewports
   respond enter **N** <R> until X at vertex (4,0) then
   enter **M** <R> – the move option
   prompt Enter new location
   enter **260,90,5** <R>
   prompt Vertex(4,0). Next/Previous ...
   enter **X** <R> – to end the Edit Vertex option
   prompt Edit Vertex/Smooth ...
   enter **X** <R> – to end the command.

8 The mesh will now be displayed as fig. (b).

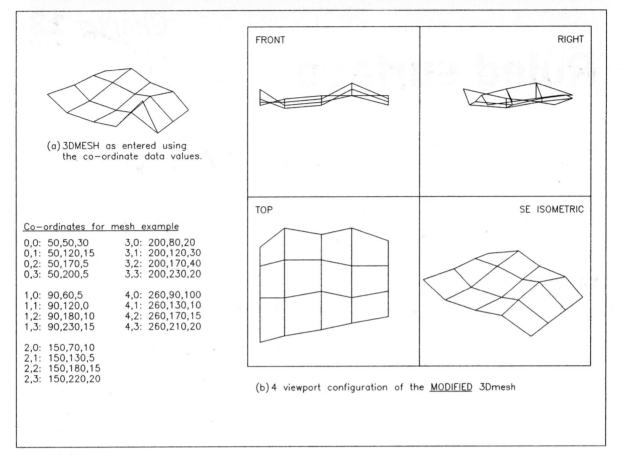

Figure 12.1 3DMESH example.

Note

1 This example is a brief introduction to the 3DMESH command.
2 The command requires the user to enter all coordinate values, and is therefore tedious to use.
3 The Edit Polyline command is used to edit 3D meshes. This is the same command as is used to edit 2D polylines.
4 There are other commands which can produce a mesh effect and we will investigate these in later chapters.
5 The 3DMESH command can be activated:
  a) from the Surfaces toolbar
  b) with **Draw–Surfaces–3DMesh** from the menu bar
  c) by entering 3DMESH at the command line.

Chapter 13
Ruled surface

A ruled surface can be created between two defined boundaries. The entities which can be used to define ruled surface boundaries are lines, arcs, circles, 2D/3D polylines and points. The surface created is a one way mesh of straight lines drawn between the two selected boundaries.

We will demonstrate the command with worked examples, the first being in 2D drawing to allow the user to become familiar with the terminology.

Example 1

1 Begin a new 2D drawing and refer to Fig. 13.1. Display the Draw, Modify, Object Snap and Surfaces toolbars.

2 Draw two lines and arcs as fig. (a), then make a new layer RULSF, colour blue and current.

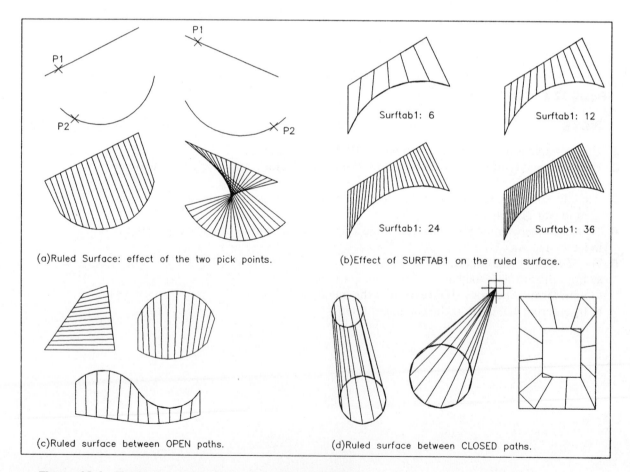

Figure 13.1 The basic ruled surface terminology and usage.

Pick points

1 Select the RULED SURFACE icon from the Surfaces toolbar and:
  prompt Select first defining curve
  respond **pick a point P1 on a line**
  prompt Select second defining curve
  respond **pick a point P2 on an arc**

2 From the menu bar select **Draw**
  Surfaces
  Ruled Surface
  prompt Select first defining curve
  respond pick a point P1 on the second line
  prompt Select second defining curve
  respond pick a point P2 on the second arc

3 The ruled surfaces added between the lines and arcs are dependent on the pick point positions of P1 and P2 – fig. (a).

Surftab1 effect

1 Refer to Fig. 13.1(b) and, using layer 0, draw a line and arc then copy them three times as shown.

2 Make layer RULSF current.

3 At the command line enter **SURTAB1** <R> and:
  prompt New value for SURFTAB1<?>
  enter 6 <R>

4 At the command line enter **RULESURF** <R> and:
  prompt Select first defining curve – pick a point on first line
  prompt Select second defining curve – pick point on first arc

5 By entering SURFTAB1 at the command line, enter new values of 12, 24 and 36 and add a ruled surface between the other lines and arcs as fig. (b).

6 The variable SURTAB1 controls the number of 'rules strips' added between the defined curves. The default is generally 6. Decide for yourself what value of SURTAB1 you want to use.

Open and closed paths

The ruled surface command can be used between:
a) two OPEN paths – lines, arcs, open polylines as fig. (c)
b) two CLOSED paths – circles, points, closed polylines as fig. (d).

The command **cannot** be used between an open and closed path, e.g. between a line and a circle. If a line and circle are selected, then the following message is displayed:

Cannot mix closed and open paths

90 *Modelling with AutoCAD*

Example 2: a 3D model

1 Open your STD3D standard sheet with layer MODEL current.

2 Refer to Fig. 13.2, and create the model base lines and arcs using the sizes given – fig. (a).

3 Make a new layer RULSRF, colour blue and current.

4 Set the SURFTAB1 variable to 18.

5 Using the Ruled Surface icon, select the following defined curves: lines 1 and 2; arcs a and b; lines v and w – fig. (b).

6 Erase the added ruled surfaces, and create the model top surface by copying the base elements:
a) from a line endpoint
b) by @0,0,50 – fig. (c).

7 With layer RULSRF still current, pick the Ruled Surface icon and select the following defined curves:
a) lines 1 and 2 : ruled surface added
b) lines 3 and 1 : ruled surface cannot be added and the following message is displayed at the command line:

Object not usable to define ruled surface – why is this?

8 When the defined curves were being selected:
a) point 3 was picked satisfactorily
b) point 1 could not be picked – you were picking the previous ruled surface added between defined curves 1 and 2.

9 Cancel the command, and erase the added ruled surface.

10 Make the following new layers:

RUL1 – red; RUL2 – yellow; RUL3 – green; RUL4 – blue

11 *a*) Make layer RUL1 current
b) Activate the ruled surface command and add a ruled surface to the base of the model – three times.

12 *a*) Make layer RUL2 current
b) freeze layer RUL1
c) add a ruled surface to the three 'outside' surfaces of the model
d) thaw layer RUL1 – fig. (d).

13 *a*) Make layer RUL3 current
b) freeze layer RUL1 and RUL2
c) ruled surface the top surface of the model.

14 *a*) Make layer RUL4 current and freeze layer RUL3.
b) Add a ruled surface to the three 'inside' surfaces of the model.
c) Thaw layers RUL1, RUL2 and RUL3 – fig. (e).

15 From the menu bar select **Tools–Hide** – fig. (f).

16 From the menu bar select **Tools**
Shade
16 Color Filled

17 Impressed with the result?

18 At the command line enter **REGEN** <R> to return the model to wire-frame representation.

19 Save the model at this stage as C:\R13MODEL\FLOWBED – it will be used later.

20 *Note*: when the ruled surface command is being used with adjacent surfaces, it is essential that layers are made for each individual surface. This allows:
  a) the surfaces to be selected as defining curves
  b) surfaced layers to be frozen
  c) colours to be added to impress.

21 **Task**: the original wire-frame model base was created from lines and arcs. It could also have been created from a single polyline/arc segment and offset to complete the base. Try this if you have time?

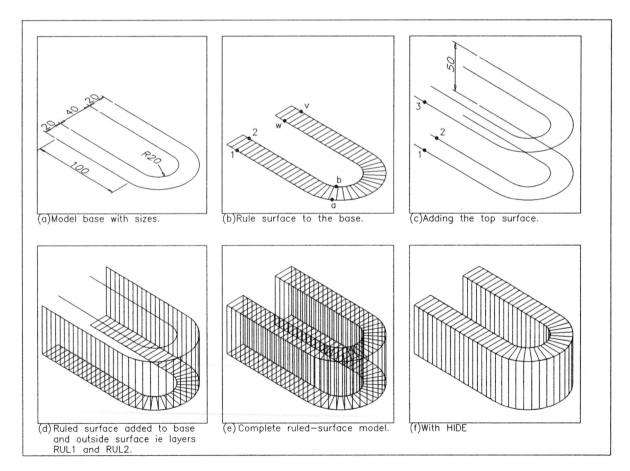

Figure 13.2 Ruled surface example 1.

Example 3

In this example we will investigate how to add a ruled surface to a model surface which has a hole in it – circular and slotted. The example will be given in 2D but the procedure is the same for a 3D model.

1 Begin a new 2D drawing and refer to Fig. 13.3.

2 Make two new layers, MODEL colour red for the outlines, and RULSRF colour blue for the ruled surface.

3 Using the LINE icon, draw a 60 unit square with a 15 radius circle at its 'centre'.

4 Using the Ruled Surface icon, pick a line of the square and the circle as the defining curves. No ruled surface is added because of the open/closed path effect – fig. (a).

5 Draw a 60 sided square with the POLYLINE icon from pt1–2–3–4–<R> as fig. (b). Rule surface? – fig. (b).

6 Draw a closed polyline square from pt1–2–3–4–c as fig. (c). This square and circle can be ruled surfaced, but the surface is not added correctly.

7 Draw a closed polyline square from pt1–2–3–4–c as fig. (d) and add a ruled surface. The added surface is not ideal as SURFTAB1 is set to 6.

8 Erase the added ruled surface and set SURFTAB1 to 24, then add a ruled surface as fig. (e).

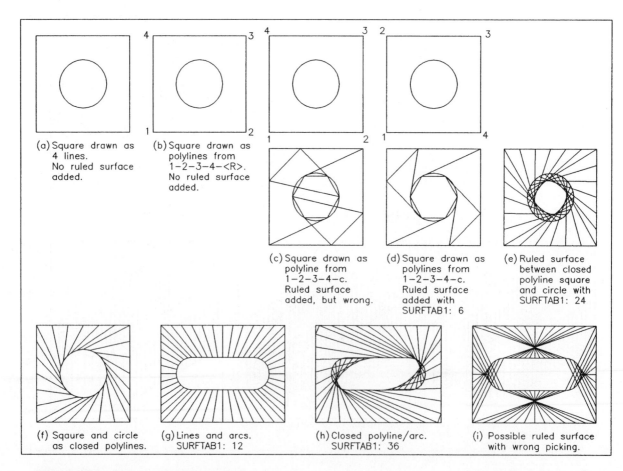

Figure 13.3 Ruled surface example 3.

9 Finally draw the square as a closed polyline and the circle as two closed polyarcs (you should be able to do this?). Add the ruled surface – fig. (f).

10 Task
a) Create the shapes in fig. (g) and fig. (h) using:
fig. (g): lines and arcs
fig. (h): closed polylines and arcs
b) Add a ruled surface:
fig. (g): use command 4 times with SURFTAB1 12
fig. (h): use command once with SURFTAB1 36
Note two of the corners?
c) The effect of picking the 'wrong way' is shown in fig. (i).

Summary

1 A ruled surface can be added between lines, circles, arcs, points and polylines.
2 The command can be activated by icon, from the menu bar or by direct keyboard entry.
3 The command can be used in 2D or 3D.
4 The command can be used between open and closed paths, but not between an open or closed path.
5 Closed paths *must be drawn in the same direction*.
6 The variable SURFTAB1 controls the ruled surface display. The default value is 6. Values of 12, 18, 24 are recommended depending on the length of the defining curves.
7 Closed polylines are the easiest to rule surface, but the final effect may not be as expected.

Activity

The ruled surface activity involves converting a wire-frame model into a surface model – *Tutorial 18*.

1 Open STD3D and create the wire-frame model using the sizes given.

2 Make several new ruled surface layers with varying colour.

3 Set SURFTAB1 to 12 or 18.

4 Set a 4 viewport configuration and set the viewpoints given.

5 Rule surface the two holes – same layer.

6 Rule surface:
a) the sides of the model on one layer
b) the horizontal and vertical surfaces with the holes
c) the base and rear

7 Display with hide and try shade in each viewport.

8 Save the model as C:\R13MODEL\RULMODEL.

Chapter 14
Tabulated surface

A tabulated surface is defined by the user selecting:

1 the **path curve** which determines the profile of the final surface.

2 the **direction vector** which determines the direction and 'depth' of the profile surface.

The path curve can be created from a line, arc, circle, ellipse, 2D or 3D polyline/arc or a spline curve. The direction vector **must be** a line or open 2D/3D polyline.

Example 1: 2D tabulated surface

1 Begin a new 2D drawing and refer to Fig. 14.1.

2 Draw pairs of lines as fig. (a) and fig. (b).

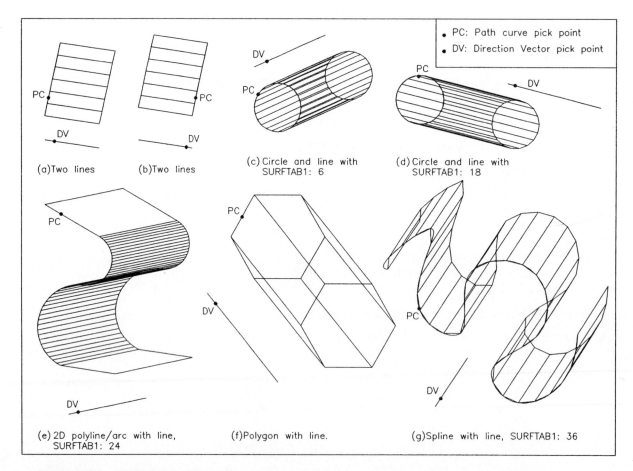

Figure 14.1 Tabulated surface in 2D.

3 From the Surfaces toolbar, select the Extruded Surface icon and:
prompt Select path curve
respond **pick the PC line indicated** in fig. (a)
prompt Select direction vector
respond **pick the DV line at the end indicated**

4 From the menu bar select **Draw**
  Surfaces
  Tabulated Surface
prompt Select path curve
respond **pick PC line indicated** in fig. (b)
prompt Select direction vector
respond **pick DV line at the end indicated**

5 A tabulated surface is added to the path curve of the two lines which were selected. The 'direction' of this tabulated surface is determined by the end of the direction vector 'picked'. This is evident in figs (a) and (b).

6 Figure 14.1 illustrates other tabulated surfaces:
  (c) circle as the path curve, line as direction vector with the system variable SURFTAB1 set to 6.
  (d) circle and line with SURFTAB1 18
  (e) 2D polyline/arc path curve and line
  (f) a polygon as the path curve and a line
  (g) a spline path curve and a line direction vector, with SURFTAB1 set to 36.

7 Try these (and other) examples for yourself.

8 *Note*: the 'definition' of the resultant tabulated surface is determined by the value of the system variable SURFTAB1.

Example 2: 3D model 1

1 Open your STD3D standard sheet with layer MODEL current and refer to Fig. 14.2

2 *a*) Create a path curve from a 2D polyline using:
From point   50,50
To point   @50,0
To point   @0,20
To point   @-20,0
To point   @0,30
To point   @50,0
To point   @0,20
To point   @-80,0
To point   close
b) Create a direction vector from a line:
From point   20,20,0
To point   20,20,80
c) Result is fig. (a).

3 Using the Extruded surface icon:
a) pick the polyline shape as the path curve
b) pick the line (at lower end) as the direction vector
c) the resultant tabulated surface is as fig. (b).

4 Use the HIDE command – fig. (c). Shade?

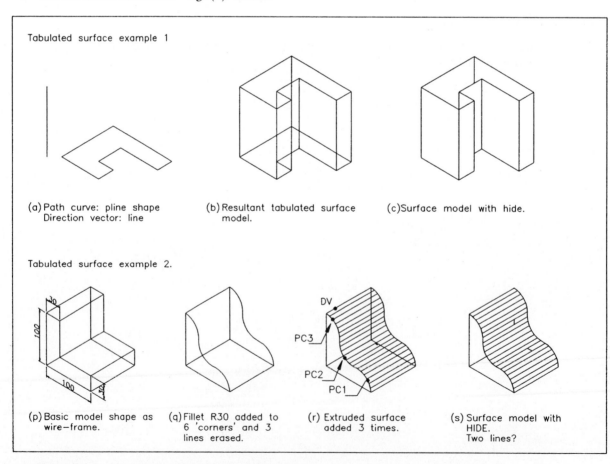

Figure 14.2 3D extruded surfaces examples.

Example 3: 3D model 2

1 Create an L-shaped wire-frame model from a 100 cube having a 'thickness' of 30 as fig. (p). The start point should be 250,50,0.

This should not give you any problems?

2 Set a fillet radius of 30, and then fillet six corners – fig. (q). You will also need to erase three lines!

3 Using the extruded surface command:
a) pick the fillet PC1 as the path curve
b) pick line DV as the direction vector
c) repeat the command for the other two fillets PC2 and PC3 using the same DV line as the direction vector
d) the result should be fig. (r).

4 Hide the model – fig. (s) – interesting effect? You should be able to reason this resultant hide model.

5 Regen to restore the wire-frame representation.

6 Save your models if required.

Summary

1 Tabulated surfaces can be used in 2D or 3D.
2 The command requires:
a) a path curve – a single entity
b) a direction vector – also a single entity.
3 The resultant surface has the same 'orientation' as the direction vector.
4 The appearance of the surface is determined by the value of the system variable SURFTAB1.

Activity

There is no formal activity with this chapter, but I would recommend that you practice some tabulated surfaces of your own. The resultant surface model can be impressive.

Chapter 15
Revolved surface

Revolved surfaces are probably the most spectacular of the surface models which can be created. The resultant model can be very complex yet creating it is very simple. The command is similar in operation to the tabulated command, the user selecting:

1 the path curve: a line, arc, circle, ellipse, 2D/2D polyshape a spline.

2 the axis of revolution: generally a line, but can be an open 2D or 3D polyshape. This is the direction vector.

Example 1: a 3D shaft

1 Open STD3D, layer MODEL current with toolbars as required. Erase the black border and set a 4 viewport configuration with:
Top left: Front   Top right: Left
Bottom left: Top   Bottom right: SE Isometric

2 Refer to Fig. 15.1. and make the bottom left (Top) viewport active.

3 *a)* Using the PLINE command, create a **CLOSED** polyline shape using the sizes given in fig. (a), the start point being 50,50. This is the path curve.
b) Create the axis of revolution by drawing a line from 0,50 to 0,150.

4 *a)* At the command line enter **SURFTAB1** <R> and set to 18.
b) At the command line enter **SURFTAB2** <R> and set to 18.

5 From the Surfaces toolbar select the Revolved Surface icon and:
prompt   Select path curve
respond   **pick the polyshape**
prompt   Select the axis of revolution
respond   **pick the line** at its 'lower' end
prompt   Start angle<0> and enter **0** <R>
prompt   Included angle (+ = ccw, – = cw)<Full circle>
enter   **360** <R> or right-click

6 A revolved surface model of the polyline is displayed.

7 Centre the model in each viewport with:
a) 0,120,0 as the centre point
b) 400 as the magnification

8 Hide the four viewports – fig. (b).

9 Save the model.

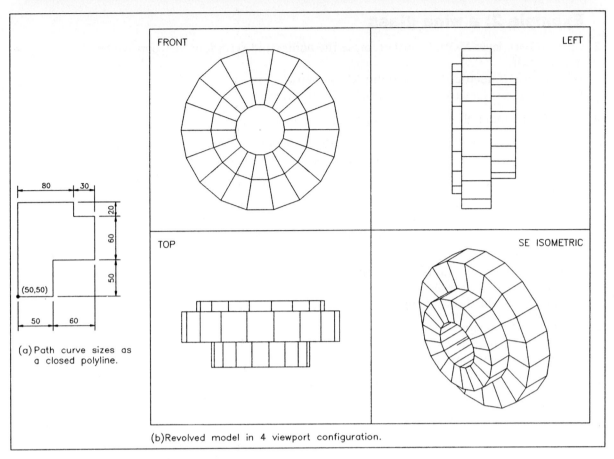

Figure 15.1 Revolved surface example 1.

Example 2: a wine glass

1. Open STD3D, layer MODEL current, erase the border and set a four viewport configuration with:
   Top left: NW Isometric   Top right: NE Isometric
   Bottom left: SE Isometric   Bottom right: SW Isometric

2. Make the bottom right viewport active and refer to Fig. 15.2.

3. *a*) With the polyline command, create an **OPEN** path curve of the wine glass using the sizes in fig. (a) as a guide. The wine glass shape is your own design.

   b) Draw in the axis of revolution as a line from (0,0) to (0,250)

4. From the menu bar select **Draw**
   Surfaces
   Revolved Surface
   a) pick the polyline as the path curve
   b) pick the line as the axis of revolution
   c) enter 0 as the start angle
   d) enter 270 as the included angle.

5. Centre each viewport:
   a) about the point 0,100,0
   b) with 200 magnification.

6. Hide each viewport – fig. (b) – then save if required.

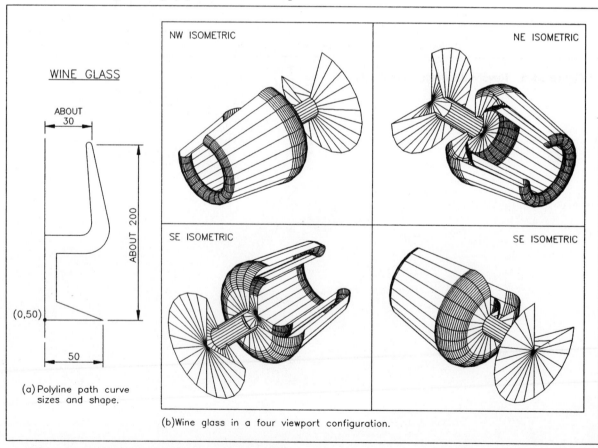

Figure 15.2 Revolved surface model example 2.

Notes

1 The two examples should demonstrate to the user that very complex surface models can be obtained with the revolved surface command.

2 The resultant surface model is a polygon mesh created with **M** and **N** facet spacing similar to a 3DMESH. The M and N facet faces are controlled by the system variables:
a) SURFTAB1: facets in the M direction (x-axis)
b) SURFTAB2: facets in the N direction (y-axis).

3 The value of surftab1 and surftab2 can greatly alter the appearance of the final surface model. At our level, a value of 18 is sufficient.

4 The start angle can vary between 0 and 360. An angle of 0 means that the surface is to begin on the current drawing plane. This is generally what is required.

5 The included angle allows the user to enter the number of degrees the path curve is to be revolved through. The 360 default gives a complete revolution, but the user can enter values between 0 and 360 to display 'cut-away' models.

6 The direction of the revolved surface is controlled by the sign of the include angle:
a) +ve for anti-clockwise revolved surface
b) −ve for a clockwise revolution.

Activity

Again I have not included any formal exercise with this chapter, but the user should practice this command. By drawing several shapes and revolving them about an axis, interesting surface models can be created. The results are dependent on your original path curve design.

Chapter 16
Edge surface

This is a surface made from a polygon mesh stretched between four **touching** edges. The edges can be lines, arcs, polylines or splines but must form a **closed loop**. The resultant mesh is determined by the value of the M and N defined facets, i.e. SURFTAB1 (M) and SURFTAB2 (N). The command can be used in 2D or 3D.

Example 1: 2D edge surfaces

1 Begin a new 2D drawing and refer to Fig. 16.1.
2 Create the following adjacent edges:
   a) four lines – fig. (a)
   b) four arcs – I used the 3 point option
   c) four single polyarcs forming a circle.

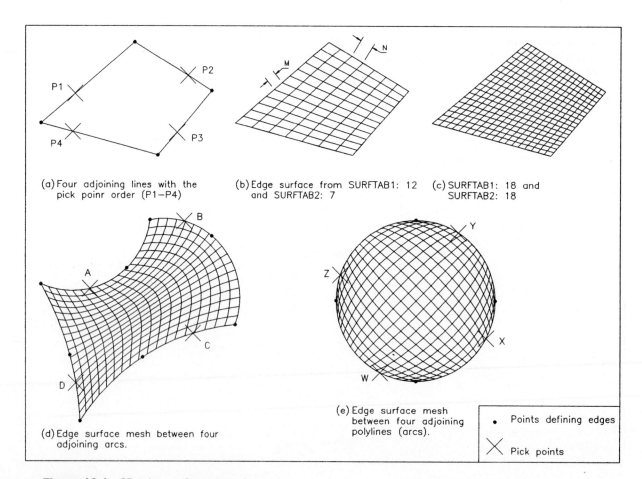

Figure 16.1   2D edge surface example.

3 At the command line enter:
  a) SURFTAB1 and set to 12
  b) SURFTAB2 and set to 7

4 Select the Edge Surface icon from the Surfaces toolbar and:
  prompt Select edge 1
  respond pick line denoted by P1
  prompt Select edge 2 and pick line denoted by P2
  prompt Select edge 3 and pick line denoted by P3
  prompt Select edge 4 and pick line denoted by P4

5 A 12 × 7 surface mesh is stretched between the four adjacent lines as fig. (b).

6 The mesh displays that the SURFTAB1 setting of 12 (M) is in the direction of P1 – the first line picked. The SURFTAB2 value of 7 (N) is in the direction of P2 – the second line picked.

7 Erase the added mesh and reset SURFTAB1 and SURFTAB2 to 18.

8 Edge surface the four lines – fig. (c).

9 From the menu bar select **Draw**
          Surfaces
              Edge Surface
  prompt Select edge 1 and pick arc denoted by A
  prompt Select edge 2 and pick arc denoted by B
  prompt Select edge 3 and pick arc denoted by C
  prompt Select edge 4 and pick arc denoted by D

10 The resultant mesh is stretched between the four arcs – fig. (d).

11 At the command line enter **EDGESURF** <R> and:
  prompt Select edge 1 and pick polyarc denoted by W
  prompt Select edge 2 and pick polyarc denoted by X
  prompt Select edge 3 and pick polyarc denoted by Y
  prompt Select edge 4 and pick polyarc denoted by X

12 The resultant mesh is interesting? – fig. (e).

13 Save the drawing if required.

Example 2: 3D edge surfaces

Three-dimensional edge surfaces are created in the same way as a 2D surfaces, but the resultant mesh has 'depth'. We will demonstrate the command by stretching an edge surface between four lines drawn in 3D, and then investigate how the added mesh can be edited.

1 Open your STD3D standard sheet and create two new layers:

OUTL: colour red and current
EDGSUR: colour blue

2 Refer to Fig. 16.2 and erase the border from your drawing.

3 Create the edges of the mesh using the LINE command with the following input:
From point   50,50,0   P1
To point   70,220,80   P2
To point   200,300,30 P3
To point   250,20,−30 P4
To point   C − fig. (a).

4 Set SURFTAB1 and SURFTAB2 to 18, and make layer EDGSUR current.

5 Using the Edge Surface command, pick the four lines to give the surface mesh – fig. (b).

Note: pick in order, i.e. P1–P2; P2–P3; P3–P4; P4–P1

6 At this stage save your mesh before proceeding to the edit section.

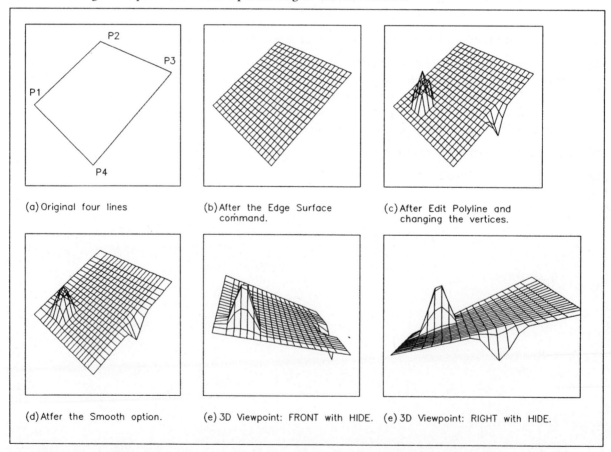

(a) Original four lines
(b) After the Edge Surface command.
(c) After Edit Polyline and changing the vertices.
(d) Atfer the Smooth option.
(e) 3D Viewpoint: FRONT with HIDE.
(e) 3D Viewpoint: RIGHT with HIDE.

Figure 16.2 3D edge surface example.

Edge surface 105

Editing a surface mesh

An edge surface is a polygon mesh and can therefore be edited. We will demonstrate the edit command using the move and smooth options, i.e. we will move several vertices of the mesh to new positions to create 'bumps and troughs', and then smooth the resultant surface.

1 From the menu bar select **Modify**
Edit Polyline

prompt	Select polyline
respond	pick the added mesh
prompt	Edit vertex/Smooth surface/ ...
enter	**E** <R> – the edit vertex option
prompt	Vertex (0,0). Next/Previous ...
and	an X appears at the (0,0) vertex
respond	enter N,U,D,L,R to move the X until the vertex at the prompt line is (3,3), i.e.
prompt	Vertex(3,3). Next/Previous ...
enter	**M** <R> – the move option of the edit vertex option
prompt	Enter new location
enter	**@0,0,40** <R>
prompt	vertex (3,3) is repositioned

Note: **Do not exit command**

2 Continue entering N,U,D,L,R then M (move option) and enter the following new vertex positions when the vertex prompt is displayed:

Vertex	Relative movement
(3,4)	@0,0,35
(3,5)	@0,0,40
(4,3)	@0,0,50
(4,4)	@0,0,80
(4,5)	@0,0,50
(5,3)	@0,0,40
(5,4)	@0,0,50
(5,5)	@0,0,55

3 When all new vertices are entered:

prompt	Vertex(5,5). Next/Previous/ ...
enter	**X** <R> – exits Edit Vertex option
enter	**X** <R> – exits the command.

4 Repeat the Edit Vertex option of the Edit Polyline command, and enter the following new relative positions for the stated vertices using the Move option:

Vertex	Relative movement
(10,18)	@0,0,−50
(9,18)	@0,0,−20
(11,18)	@0,0,−20
(11,17)	@0,0,−10
(10,17)	@0,0,−30
(9,17)	@0,0,−10

5 The mesh after these vertex moves appears as fig. (c).

106 *Modelling with AutoCAD*

6 From the menu bar select **Modify**
Edit Polyline

prompt Select polyline
respond pick the mesh
prompt Edit vertex/ ...
enter **S** <R> – the smooth option

7 The polygon mesh surface is 'smoothed' – fig. (d).

8 Use the 3D Viewpoint Presets to view the surface:
a) from the Front – fig. (e)
b) from the Right – fig. (f).

9 At the command line enter **SURFTYPE** <R>
prompt New value for SURFTYPE<6>
enter **5** <R>

10 The edge surface will be drawn as a Quadratic B-spline curve.

11 Save the resultant surface if required.

Example 3: 3D edge surface using splines

The previous example demonstrated how an edge surface can be added between four adjacent lines. The real benefit of the command can be seen when an edge surface is added between curvilinear edges, and we will demonstrate this with four spline edges.

1 Open STD3D, erase the border and set a four viewport configuration with:

Top left: NE Isometric Top right: NW Isometric

Bottom left: Top Bottom right: SE Isometric

2 Make layer MODEL current and the bottom left (Top) viewport active.

3 Using the SPLINE command, draw four spline curves using the following coordinate input, picking start and end tangent points to suit:

spline1	spline2	spline3	spline4
50,50,0	80,240,20	280,240,0	330,45,0
80,120,40	160,220,80	280,150,50	150,40,80
60,180,80	280,240,0	330,45,0	50,50,0
80,240,20	right-click	right-click	right-click
right-click			

4 Make a new layer EDGSUR, colour blue and current.

5 Using the Edge Surface icon, pick the four spline lines 'in order' to add the surface mesh between the four spline curves.

6 The effect is displayed in Fig. 16.3. with HIDE.

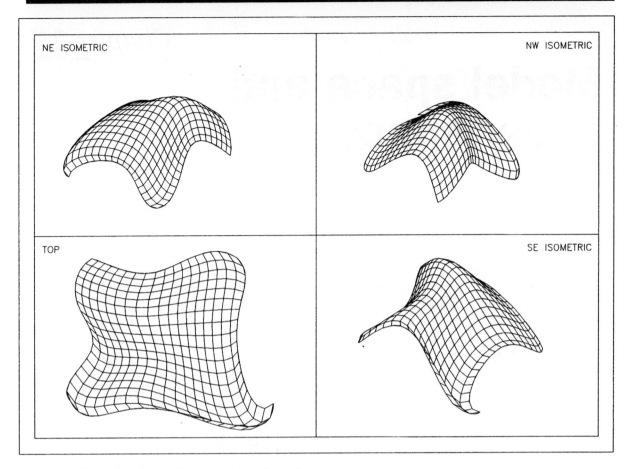

Figure 16.3 3D edge surface example using splines.

Summary

1 An edge surface is a polygon mesh added between four adjoining lines, arcs, polylines and splines.
2 The four edges must be touching at their ends.
3 The resultant edge surface can be edited using the polyline edit (PEDIT) command.
4 The surface added is called a **COONS** patch and is bicubic, i.e. one curve is in the M direction and one curve is in the N direction.
5 The surface appearance can be altered by the SURFTYPE variable and:

SURFTYPE 5 – Quadratic B-spline
SURFTYPE 6 – Cubic B-spline
SURFTYPE 8 – Bezier curve

6 The surface mesh density of an edge surface is controlled by two variables: SURFU – controls the density in the M direction (default 6) SURFV – controls the density in the N direction (default 6)

Chapter **17**

Model space and paper space

AutoCAD has multi-view capabilities which allow the user to layout, organize and plot multiple views of a single drawing. Multiple views are generally used with 3D drawings but can also be used in 2D. The multiple viewport concept has already been used in the creation of our wire-frame and surface models. The viewports which we will now discuss are used in exactly the same way as these viewports, although their creation is slightly different. To fully understand multi-views, the user must be familiar with the terms **model space** and **paper space** – two entirely different drawing environments.

Model space

This is the drawing environment that exists in any viewport and when a new drawing is started, i.e. it is the default 'setting'. All drawings which you have so far created have been completed in model space. Model space is used for draughting and design work and the setting of viewpoints. There is however one major disadvantage when working in model space – only the current viewport can be plotted, i.e. multiple viewports cannot be plotted on the one sheet of paper.

Paper space

This is a drawing environment which is independent of model space. It is primarily used to layout and arrange viewports, but can also be used to add entities to a drawing – particularly 2D text. In paper space the 3D viewpoint commands are not valid, but the real benefit of paper space is that all viewports can be plotted on a single sheet of paper.

Tilemode

AutoCAD allows two different types of viewport to be created. These were briefly discussed in Chapter 7 and are:

a) Tiled or fixed viewports. These are created in model space and have been used in all our previous work. Once created, the viewports cannot be moved. These viewports are simply called VIEWPORTS.
b) Untiled or floating viewports. These are paper space viewports and can be re-positioned. New viewports can also be added to the drawing at any time. These viewports are generally referred to as MULTI-VIEWPORTS.

The viewport 'type' is determined by a system variable (**TILEMODE**) and:

a) TILEMODE 1: model space (fixed) viewports
b) TILEMODE 0: paper space (floating) viewports.

Tiled (model space) viewports are always displayed as edge-to-edge and fill the screen like a tiled wall. Untiled (paper space) viewports can be positioned anywhere on the screen area with spaces between them if required. They are can also be copied, stretched and re-sized.

Toggling between paper space and model space

All readers should be familiar with the toggle concept, e.g. toggling the grid on/off with the F7 key or from the status line with GRID. It is possible to toggle between paper space and model space but only if the TILEMODE variable is set to 0.

The toggle effect can be activated using:

1 *command line*
   a) if in paper space, toggle to model space with **MS** <R>
   b) if in model space, toggle to paper space with **PS** <R>

2 *status bar*
   a) double left-click on PAPER to toggle to paper space
   b) double left-click on MODEL to toggle to model space.

Model/paper space example

This example will be in 2D, but the concepts are the same for a 3D drawing. For users who are new to the model/paper space idea, this 2D example will probably be easier to understand.

1 Begin a new 2D drawing and set the following layers:
BORDER: colour white
MODEL: colour red and current
VP: colour yellow

2 Refer to Fig. 17.1 and draw a few entities as fig. (a). Also draw a rectangular drawing border:
From 0, 0; to @380, 0; to @0, 270; to @−380, 0; to close.

3 From the menu bar select **View** and note:
   a) tick at Tiled Model Space, i.e. active
   b) Floating Viewports not active – in grey
   c) all other options are active – bold type
   d) cancel the pull-down.

4 Move the cursor cross-hairs down into the status line area and:
   a) position the arrow over the word MODEL
   b) double left-click

   and:

   a) drawing disappears – don't panic
   b) paper space icon displayed
   c) command line displays:
      tilemode
      New value for TILEMODE<1>: 0
      Regenerating drawing.

5 We have now entered the paper space environment.

110 *Modelling with AutoCAD*

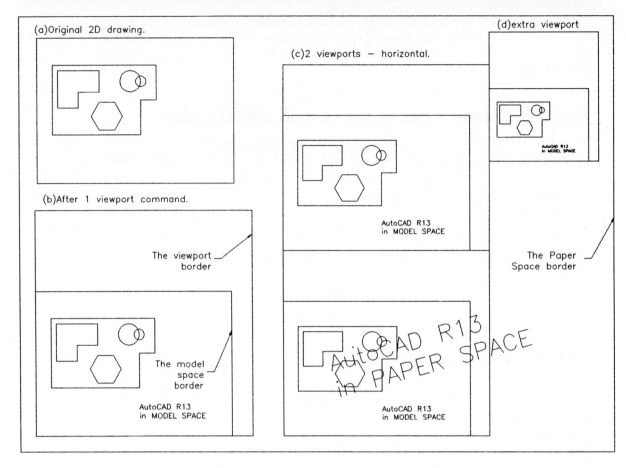

Figure 17.1 Model/paper space example.

6 Make the VP (yellow) layer current.

7 From the menu bar select **View** and note:
   a) tick at Paper space – now active
   b) Floating Viewports active – bold type
   c) Tiled Viewports not active
   d) the 3D View ... commands are not active.

8 From this pull-down menu select **Floating Viewports**
                                    1 Viewport

   prompt   ON/OFF ...<First Point>
   enter   **10,10** <R>
   prompt   Other corner
   enter   **150,150** <R>

9 A yellow viewport border will be displayed with the red drawing inside it.

10 At the command line, enter **MS** <R> and:
   a) you enter the model space environment
   b) cross-hairs contained within the yellow viewport – moving out of the viewport gives an arrow
   c) Zoom-All to give fig. (b).

11 At the command line, enter **PS** <R> to return to the paper space environment.

Model space and paper space **111**

12 Make layer BORDER (white) current and draw a paper border with the line command:

From 0, 0; to @380, 0; to @0, 270; to @–380, 0; to close.

This is a paper space border and represents the 'boundary' of our A3 paper, i.e. all work will be contained within this rectangular drawing area.

13 Make layer VP current.

14 From the menu bar select **View**
       Floating Viewports
       2 Viewports
prompt Horizontal/<Vertical>
enter   **H** <R> – horizontal option
prompt Fit/<First point>
enter   **170,10** <R>
prompt Second point and enter **300,240** <R>

15 Two new viewports will be displayed with the red drawing in each.

16 Enter model space with **MS** <R>, make each new viewport active and zoom-all. This gives fig. (c).

17 Return to paper space and make VP the current layer.

18 From the menu bar select View–Floating Viewports–1 Viewport to create a new viewport with:
a) 370, 260 as the first point
b) 300, 180 as the other corner.

19 Enter model space, make the new viewport active and zoom-all to give fig. (d).

20 Make a new layer, TEXT colour blue and current.

21 Still in model space, make the first created viewport active and using the DTEXT icon:
a) enter 200,50 as the start point
b) enter a height of 10 and rotation of 0
c) enter the text as: AutoCAD R13<R>
       in MODEL SPACE<R>

22 The text is added to the other viewports.

23 Enter paper space with PS <R>

24 Using the DTEXT icon:
a) enter 200,50 as the start point
b) enter 10 as the height and 15 as the rotation – to differ this text from the previous text
c) enter the text as: AutoCAD R13<R>
       in PAPER SPACE<R>

25 Redraw and your drawing should resemble Fig.17.1 – less fig. (a).

26 *Note*
a) Text added in model space is displayed in all viewports.
b) Text added in paper space only applies to paper space.

27 While still in paper space, save your drawing and exit AutoCAD.

112 *Modelling with AutoCAD*

Working with paper space viewports

Paper space (or floating) viewports are used in exactly the same way that the tiled viewports were used, i.e. we can enter model space and set viewpoints as required – this will be demonstrated in the next chapter with 3D models. For our 2D drawing we will investigate how the floating viewports can be modified.

1 Open the drawing saved previously – you should still be in paper space.

2 Enter model space and make the first created viewport active.

Zoom command in paper space

1 Zoom in an one of the red entities.

2 The shape is enlarged in the active viewport only, but the drawing layout is unaltered.

3 Zoom–Previous then enter paper space.

4 From the menu bar select View–Zoom–Window and window the first viewport created using the yellow border as the window.

5 The complete viewport is enlarged on the screen.

6 You can now enter model space and work on the 'larger size' drawing if required.

7 In paper space, zoom-previous to return the complete drawing.

Modifying paper space viewports

Unlike tiled (model space) viewports, floating (paper space) viewports can be repositioned anywhere on the screen.

1 Paper space active and refer to Fig. 17.2.

2 Select the MOVE icon and:
> *prompt*   Select objects
> *respond* **pick the yellow border of the first created viewport**
>   then right-click
> *prompt*   Base point ...
> *respond* **INTersection and pick lower left corner of viewport**
> *prompt*   Second point ...
> *enter*   **10,115** <R> – fig. (a).

3 Using the move icon again:
> *a*) pick yellow border of smallest viewport then right-click
> *b*) enter 300,180 as the base point
> *c*) enter 10,20 as the second point – fig. (b).

4 Now move the two joining viewports from any point by **@–10,15** to give fig. (c).

5 With the COPY icon:
> *a*) pick the smallest (moved) viewport border
> *b*) enter 10,20 as the base point
> *c*) enter 300,120 as the other point – fig. (d)

6 Drawing layout as Fig. 17.2

7 *Note*:
> *a*) all entities drawn within a viewport are moved, copied
> *b*) the paper space text has not changed.

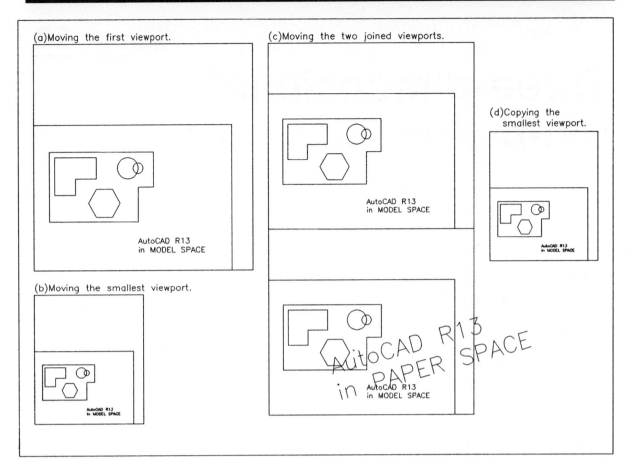

Figure 17.2 Modifying floating (paper space) viewports.

Why use floating (untiled) viewports?

New users to floating viewports may be unconvinced of their advantages over tiled viewports, and I have often been asked 'why use this paper space idea when tiled viewports allow simultaneous work on models?'. The following table gives a comparison between the two environments:

Model space
model created
model can be modified
tiled (fixed) viewports
tilemode: 1
plot only current viewport
viewports restricted in size
viewports 'fill screen'
viewports cannot be altered
3D commands permitted
WCS or UCS icon

Paper space
no model created
no modification allowed
untiled (floating) viewports
tilemode: 0
all viewports can be plotted
viewports can be made to any size
viewports user specific
viewports can be moved, copied, etc.
no 3D commands allowed
paper space icon

Chapter **18**

Three-dimensional multi-view drawings

The previous chapter introduced the model and paper space concept of floating (untiled) viewports as well as the TILEMODE variable. The example used to investigate the topic was in 2D but untiled viewports are not really a 2D drawing aid. Their main advantage is with 3D models as they allow the plotting of multiple views of a component on a single sheet of paper.

We will investigate the 'power' of untiled viewports with three examples, and discuss how to:

a) create floating viewports for wire-frame and surface models which have already been created and saved – hopefully!
b) alter viewports configurations
c) create a new model using viewports which are pre-set
d) centre models in viewports
e) use viewport specific layers.

Note: two of the examples used in this chapter should have already been created and saved. If they are not saved, then it will be necessary for you to return to the appropriate chapter and create the model.

Example 1: WORKDRG

The WORKDRG wire-frame model has been used in several chapters, and will now be used as our first 3D multi-view exercise.

Getting ready

1 Open your wire-frame model WORKDRG which should be displayed with text, hatching and dimensions?

2 Restore the UCS BASE.

3 *a*) make a new layer called VP colour yellow
 b) freeze the dimension, text and hatch layers
 c) erase the black border if displayed.

4 Make layer OBJECTS (blue) current.

5 Set SURFTAB1 to 18 and using the Ruled Surface icon pick the two blue circles.

6 Set SURFTAB1 to 12 and rule surface the two triangles, picking the appropriate corresponding lines – you will probably have to zoom in on the triangles to pick the lines. Remember to zoom out again.

Creating the viewports

1 Make the new layer VP (yellow) current.

2 Double left-click on MODEL in the status bar – now entered paper space.

3 From the menu bar select **View–Floating Viewports–4 Viewports**
prompt Fit/<First point> and enter **10,10** <R>
prompt Second point and enter **370,260** <R>

4 Four yellow viewports will be created with the wire-frame model displayed in each.

Making the A3 drawing sheet

1 Make layer 0 current

2 Using the LINE command, create a border effect:

from 0, 0; to @380, 0; to @0, 270; to @–380, 0; to close

3 This border 'encloses' the four yellow viewports.

4 So far all work has been in paper space.

Setting the viewpoints

1 Enter model space with a double left-click on PAPER in the status bar.

2 Make the top left viewport active.

3 From the menu bar select **View–3D Viewpoint Presets–Front**

4 Making the other viewpoints active, use the 3D Viewpoint Presets to set the following viewpoints:
a) bottom left: Top
b) top right: Left
c) bottom right: SE Isometric

5 The wire-frame model will be displayed at the set viewpoint in the viewports, but the 'scale' in each viewport is different.

Centring the model in the viewports

In this exercise we will use the zoom–centre method of centring a model in a viewport. The basic wire-frame model is cuboid in shape, the overall sizes being $150 \times 80 \times 100$. The centre point of the model is thus at (75, 40, 50) – these figures being relative to the UCS BASE position. Now you know why we restored this earlier.

1 Make the bottom left (top) viewport active.

2 From the menu bar select **View–Zoom–Center** and:
prompt Center point and enter **75,40,50** <R>
prompt Magnification or height <??> and enter **125** <R>

3 The top view of the model will be centred in the viewport.

4 In the other viewports zoom-centre:
a) about the point 75,40,50
b) 125 magnification, but 200 in the 3D viewport

Adding the final touches

1 Enter paper space by entering **PS** <R> at the command line.

2 Make layer 0 current.

3 Using the DTEXT icon add the following items of text at height 5 and 0 rotation:
 a) start point 15, 120, text TOP
 b) right justified at 180, 245, text FRONT
 c) start point 200, 145, text LEFT
 d) right justified at 365, 120, text SE ISOMETRIC

4 Also add the following item of text:
 a) start point 10,262
 b) height 5, rotation 0
 c) text item: WORKDRG with multi-viewports created in PAPER SPACE.

5 Your first paper space multi-view drawing is complete, and should resemble Fig. 18.1.

6 Save your model, but not as WORKDRG.

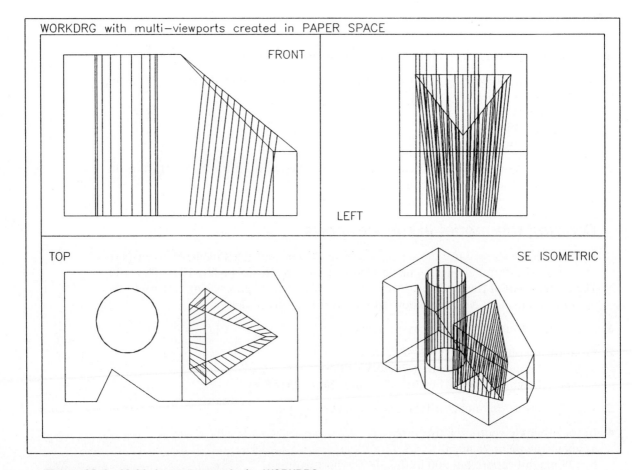

Figure 18.1 Multi-viewport example 1 – WORKDRG.

Example 2: FLOWBED

This example will use a ruled surface model to demonstrate how several viewports can be created in paper space. The flower bed model was created in Chapter 13. The steps in creating the final drawing sheet layout are similar to the first example, so the steps are given as a list of operations, and not divided in sections as example 1.

1 Open the saved ruled surface model FLOWBED and erase all entities but the actual model.

2 Enter paper space with a double left-click on MODEL in the status bar.

3 Make a new layer: VP, colour yellow and current.

4 From the menu bar select **View–Floating Viewports–2 Viewports** and:
    prompt   Horizontal/<Vertical>
    enter   **H** <R>
    prompt   Fit/<First point> and enter **10,10** <R>
    prompt   Second point and enter **200,250** <R>

5 Two horizontal yellow viewports with the model in each are displayed.

6 Repeat the View–Floating Viewports–2 Viewports and create two more horizontal viewports using:
    a) 370,10 as the first point
    b) 260,130 as the second point

7 Using the 1 Viewport option, create two single viewports with:
    a) 200,105 as the first point
        260,175 as the second point
    b) 275,150 as the first point
        355,225 as the second point

8 You have now created six viewports, each displaying the ruled surface model of FLOWBED.

9 With layer 0 current, draw an A3 sized border around the six yellow viewports using the LINE icon and the same coordinate input as before:

    From 0, 0; to @380, 0; to @0, 270; to @−380, 0; to close.

10 Enter model space – **MS** <R>

11 Set a UCS–BASE to the point 50,50,0. This should have been the start point of your original model. Check it out and readjust your coordinates if required. Remember UCSICON–All–ORigin?

12 Refer to Fig. 18.2 and use the 3D Viewpoint Presets to set the required viewpoint in each of the viewports.

13 We now need to centre the model in each viewports and will use the **XP** option. As the viewports are not all the same size, different XP values will be required.

15 Make the lower left (top) viewport active and at the command line enter **ZOOM** <R>
    prompt   All/Center ...
    enter   **1XP** <R>

16 The ruled surface model will be centred in the selected viewport.

17 Using the ZOOM–?XP entry from the command line, refer to Fig. 18.2 and centre the model in each viewport using the XP value given.

18 Make layer TEXT (green) current and using the DTEXT icon, add the following item of text:
   a) start point 0,–12
   b) height 8, rotation 0
   c) item: MODEL SPACE CREATED

19 The text item will be displayed in all six viewports

20 Enter paper space, PS <R>, and make layer 0 current.

21 Using the DTEXT icon add the following text item:
   a) start point 50,60; height 8; rotation 15
   b) text: THIS TEXT HAS BEEN ADDED IN PAPER SPACE.

22 A3 layout now as Fig. 18.2 and can be saved.

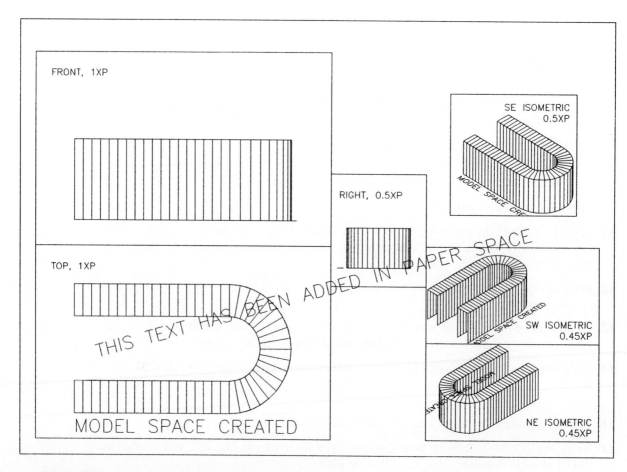

Figure 18.2 multi-view example 2 – ruled surface model.

Three-dimensional multi-view drawings **119**

Example 3: a new 3D model

The two previous examples used models which were already created, and the viewport layout was made after the model. In this example we will create a basic multi-view A3 layout, save it for future use, and then create the model.

The A3 sheet

1 Open your STD3D standard sheet which has settings and layers (DIMS, HATCH, MODEL, OBJECTS, TEXT) with a viewpoint of SE Isometric.

2 Erase the black border.

3 Make two new layers:
a) VP, colour yellow and current
b) A3SHEET, colour white (black).

4 Enter paper space with a double left-click on MODEL.

5 Use View–Floating Viewpoints to create a 4 viewport setup with:
a) first point: 10,10
b) second point: 370,260

6 With layer A3SHEET current, use the line command to draw an A3 border as before using:

From 0, 0; to @380, 0; to @0, 270; to @–380, 0; to close.

7 With layer 0 still current, add the following items of text with height 4 and 0 rotation:
a) at 15,250: FRONT
b) at 15,15: TOP
c) right justified at 365,250: LEFT
d) right justified at 365,15: 3D VIEW.

8 Add another item of text:
a) height 6, rotation 0
b) at 10,262
c) text item: TITLE.

9 Enter model space (MS <R>) and use the 3D Viewpoint Presets to set the viewpoints in the following viewports:
Top left: Front   Top right: Left
Bottom left: Top Bottom right: SE Isometric

10 At the command line enter **UCSICON** <R>
prompt ON/OFF ...
enter   **A** <R> – for all viewports
prompt On/OFF ...
enter   **OR** – for origin

11 Make the lower right (3D) viewport active and layer MODEL current.

12 We now have an A3 prototype drawing set for multi-view use. This drawing will be used for all future multi-view work – especially solid modelling.
Note: *a*)   Viewpoints can be altered to suit model.
       b)   Still need to set UCS's dependent on the model.
       c)   The viewports have not yet been 'centred'. This also depends on the model being created.

13 At this stage save the A3 sheet layout as **C:\R13MODEL\A3MVPROT**.

14 Continue with the worked example.

120 *Modelling with AutoCAD*

The new model

As a change from the two completed examples, we will create a 3D model using the 3D objects – which never seem to be used but are very powerful. As we create a new 3D object, we will change its colour.

1 In model space with layer MODEL current and the lower right (3D) viewport active?

2 Display the Surfaces toolbar and any others of your choice.

3 Select the Box icon from the Surface toolbar and:
prompt   Initializing ... 3D Objects loaded
then   Corner of box
enter   **100,100,0** <R>
prompt   Length and enter **25** <R>
prompt   Cube/<Width> and enter **C** <R> – the cube option
prompt   Rotation ... and enter **15** <R>

4 A red cube is displayed in the four viewports.

5 Repeat the box icon selection and:
prompt   Corner of box and enter **50,50,0** <R>
prompt   Length and enter **150** <R>
prompt   Cube/<Width> and enter **150** <R>
prompt   Height and enter **100** <R>

6 Using the Properties icon from the Object Properties toolbar, change the colour of this box to blue.

7 Now add the following four additional 3D objects using the information supplied:
  a) *Wedge*
Corner	50,200,0
Length	50
Width	150
Height	100
Rotation	180
Colour	green

  b) *Cone*
Base circle	125,125,100
Base radius	75
Top radius	0
Height	100
Segments	16
Colour	magenta

  c) *Dish*
Centre	125,125,0
Radius	75
Long segments	16
Lat segments	8
Colour	cyan

  d) *Torus*
Centre	125,125,150
Torus radius	100
Tube radius	20
Torus segments	16
Tube segments	16
Colour	red

Three-dimensional multi-view drawings **121**

8 The model is now complete and displayed in the four viewports, but has not yet been 'centred'.

9 Set and save a UCS BASE at the point 50,50,0 – the model start point. Icon at this point?

10 Using the zoom-centre method, centre each viewport using:
a) 50,75,65 as the centre point – why these?
b) 300 magnification.

11 Enter paper space with **PS** <R>

12 From the menu bar select **Modify**
                        Edit Text ...
prompt	<Select an annotation ...
respond	**pick the TITLE text**
prompt	Edit Text dialogue box
with	TITLE as the text
enter	**Multi-view model created from 3D OBJECTS**
then	pick OK
prompt	<Select an annotation ...
respond	right-click

13 At this stage save your layout as **MVMOD** as it will be used during the chapter on dynamic viewing.

Adding two more viewports

1 You should still be in paper space, so make layer VP current.

2 Using View–Floating Viewports, make two new viewports with:
   a) from 190,135 to 240,190
   b) from 380,190 to 330,135

3 In model space, set new viewport *a*) to NE Isometric and new viewport *b*) to SW Isometric.

4 Zoom–centre these new viewports about the point 50, 75, 65 at 300 magnification.

5 At this stage your A3 layout should resemble Fig. 18.3.

6 Hide each viewport – no red cube is seen.

7 Enter **REGENALL** <R> at the command line to return the model to wire-frame representation.

8 Questions
   a) Why did I position one of the viewports 'outside' the existing yellow viewport border?
   b) Why is the small viewport in the 'centre' of the screen wrongly placed?

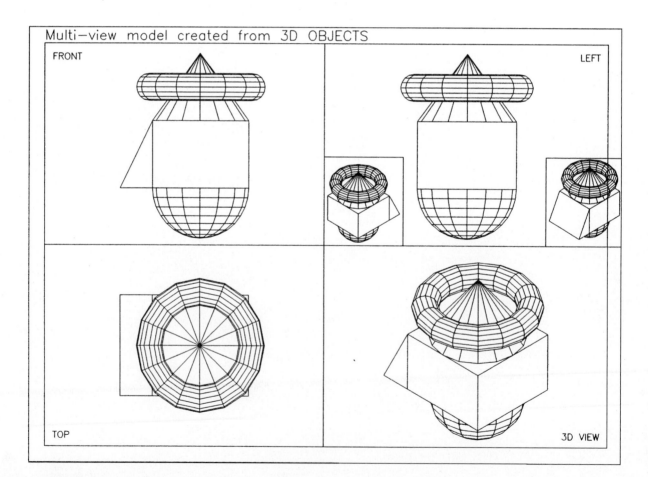

Figure 18.3 Multi-view example 3.

Three-dimensional multi-view drawings **123**

Adding dimensions to the model

1 In model space with the 3D viewport active, erase the red torus and the cyan dish then REDRAWALL.

2 Make the lower left viewport active and enter paper space.

3 In paper space use zoom-window on the lower left viewport then return to model space.

4 Make layer DIMS (magenta) current.

5 Add two linear dimensions – one horizontal and one vertical.

6 Enter paper space, zoom-previous, then return to model space.

7 The two dimensions will be displayed in the other viewports.

8 Make the top left viewport active

9 Select from the menu bar **Data**
Layers
prompt   Layer Control dialogue box
respond   1 pick DIMS layer line – turns blue
           2 pick **CUR VP: Frz**
           3 Note warning at bottom of box
           4 pick OK

10 The two magenta dimensions in the top viewport disappear?

11 Repeat step 7 in the two right viewports.

12 Dimensions should now only be displayed in the lower left viewport.

13 *Note*: the steps in this section are an introduction to viewport specific layers. The topic will be discussed in greater detail in a later chapter.

14 As this completes the third example, exit the drawing **without** saving the changes.

Summary

1 Multiple viewports can be created in paper space.
2 Paper space viewports allow prototype drawings to be created for different sized sheets.
3 Paper space viewports are used in exactly the same way as tiled viewports were used, i.e. viewpoints, zoom effect, etc.
4 Paper space viewports are particularly suited to 3D modelling.
5 Paper space viewports use viewport specific layers.
6 Viewports can be centred with:
    a) zoom–centre and a magnification value
    b) zoom with an XP value but different sized viewports require different XP values.
7 Any future multi-view work should use the A3MVPROT drawing, saved with the four viewport configuration. This can easily be altered to suit individual requirements.

Chapter 19

Three-dimensional geometry

There are three commands specific to 3D modelling, these being 3D Rotate, 3D Mirror and 3D Array. In this chapter we will investigate these commands as well as how to use the Align command with 3D models.

Getting started

1 Open your STD3D standard sheet (not A3MVPROT) with layer MODEL current and refer to Fig. 19.1

2 Using the Surfaces toolbar, create a box:
   a) corner at 50,50,0
   b) length 100; width 100; height 40; rotation 0
   c) colour: red

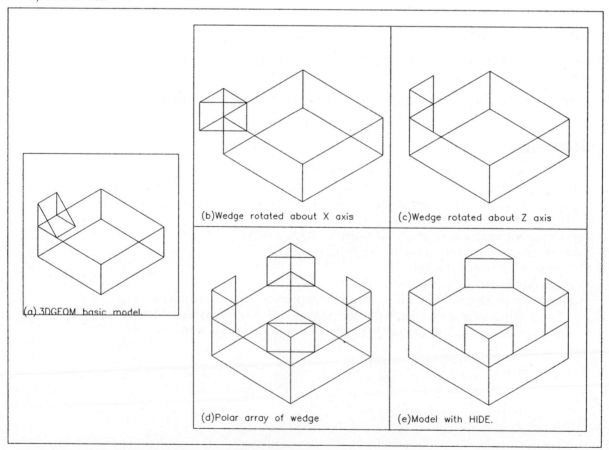

Figure 19.1 3DGEOM model for the 3D commands.

3 Using the Surfaces toolbar, create a wedge:
   a) corner at 50,50,40
   b) length 30; width 30; height 30; rotation 0
   c) colour: blue

4 Set the UCS origin at 50, 50, 0 and save as BASE

5 At this stage your model should resemble fig. (a).

6 Save as **3DGEOM**.

3D rotate

Using Modify–Rotate results in a 2D command, i.e. the selected entities are rotated in the current *XY*-plane. Objects can be rotated in 3D relative to the *X*, *Y* and *X* axes with the 3D Rotate command. We will demonstrate the command by:
a) rotating the blue wedge
b) rotating the resultant model

Rotating the wedge

1 From the menu bar select **Construct–3D Rotate**
   prompt Select objects
   respond **pick the blue wedge** then right-click
   prompt Axis by Object/Last ...
   enter   **X** <R> – the *X*-axis option
   prompt Point on *X*-axis<0,0,0>
   enter   **0, 0, 40** <R> – why these coordinates?
   prompt <Rotation angle> ...
   enter   **90** <R>

2 The wedge is rotated 'out of the box' about the *X*-axis – fig. (b).

3 Select the 3D Rotate icon from the COPY flyout of the Modify toolbar and:
   prompt Select objects
   respond pick the blue wedge then right-click
   prompt Axis by Object ...
   enter   **Z** <R> – the *Z*-axis option
   prompt Point on *Z* axis<0,0,0>
   enter   **0,0,40** <R>
   prompt <Rotation angle> ...
   enter   **90** <R>

4 The blue wedge is now aligned as required – fig. (c).

5 Select the Polar Array icon (a 2D command) and:
   prompt Select objects and pick the blue wedge then right-click
   prompt Center point of array and enter 50,50
   prompt Number of items and enter 4
   prompt Angle to fill and enter 360
   prompt Rotate objects and enter Y

6 The wedge is arrayed to the four top corners of the box – fig. (d).

7 As this model will be used with the other commands, save it at this stage as **3DGEOM**.

8 Try the following:
   a) Tools–Hide – fig. (e)
   b) Tools–Hide–16 Color Filled – interesting?
   c) enter **REGEN** <R> to return the model to wire-frame representation

Rotating the model

1 Ensure 3DGEOM model is displayed and refer to Fig. 19.2

2 From the menu bar select **Construct–3D Rotate**
prompt Select objects
respond **window the model** then right-click
prompt Axis by Object ...<2 points>
respond **INTersection and pick pt1**
prompt 2nd point on axis
respond **INTersection and pick pt2**
prompt <Rotation angle ...
enter 90 <R>

3 The model is rotated about the selected points – fig. (b) and is displayed with hide – fig. (c).

4 Activate the 3D Rotate command again and:
prompt Select objects
respond window the model then right-click
prompt Axis by Objects ...
enter Z <R> – the Z axis option
prompt Point on Z axis<0,0,0>
enter 0,0,0 <R>
prompt <Rotation angle ...
enter –90 <R>

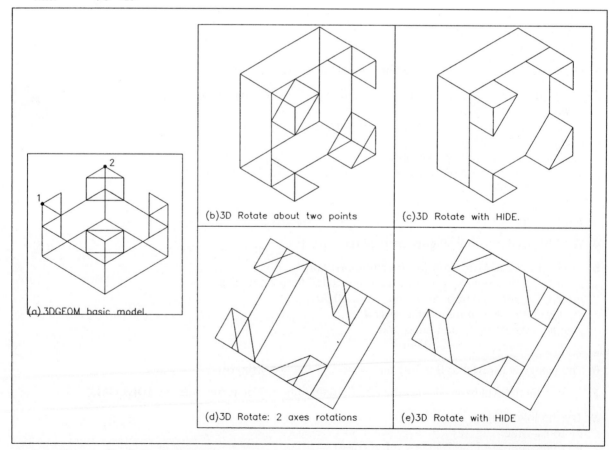

Figure 19.2 3DGEOM model – 3D rotate.

5 Repeat the 3D Rotate command and:
  a) enter **P** <R> to select the model then right-click
  b) enter **X** for the axis option
  c) enter **0,0,0** as point on axis
  d) enter **−45** as the rotation angle

6 The model is re-orientated as fig. (d) and with hide – fig. (e).

7 This completes the 3D Rotate investigation. Save?

3D Mirror

This command allows models to be mirrored about selected points or about the three X–Y–Z planes.

1 Open model 3DGEOM and refer to Fig. 19.3 – UCS BASE?

2 From the menu bar select **Construct–3D Mirror**
  prompt   Select objects
  respond   **window the model** then right-click
  prompt   Plane by Object/Last ...
  enter   **XY** <R> – the XY plane option
  prompt   Point on XY plane<0,0,0>
  respond   right-click
  prompt   Delete old objects <N>?
  enter   **Y** <R>

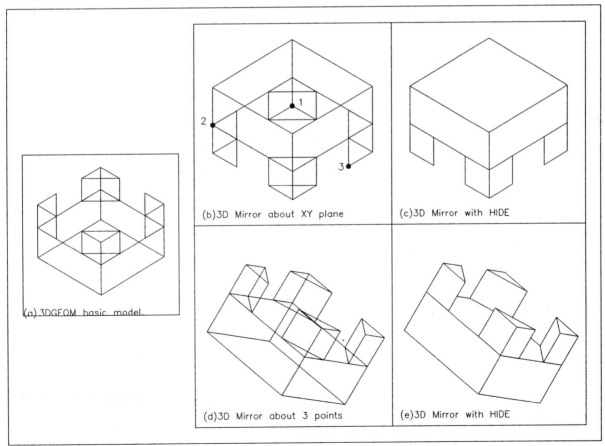

Figure 19.3 3DGEOM model – 3D mirror.

128 *Modelling with AutoCAD*

3 The model is 'flipped over' – fig. (b) and with hide – fig. (c).

4 Select the 3D Mirror icon from the Copy flyout of the Modify toolbar and:
    prompt   Select objects
    respond   window the model then right-click
    prompt   Plane by Object ...<3 points>
    respond   **INTersection and pick pt1**
    prompt   2nd point on plane
    respond   **INTersection and pick pt2**
    prompt   3rd point on plane
    respond   **INTersection and pick pt3**
    prompt   Delete old objects<N>?
    enter   **Y <R>**

5 The model is mirrored about the three selected points – fig. (d) and is displayed with hide – fig. (e).

6 Save if required?

3D Array

The 3D Array command is similar in operation to 2D Array. Both rectangular and polar arrays are possible, the rectangular array having rows and columns as well as levels in the Z-direction. The result of the polar command requires some thought with 3D.

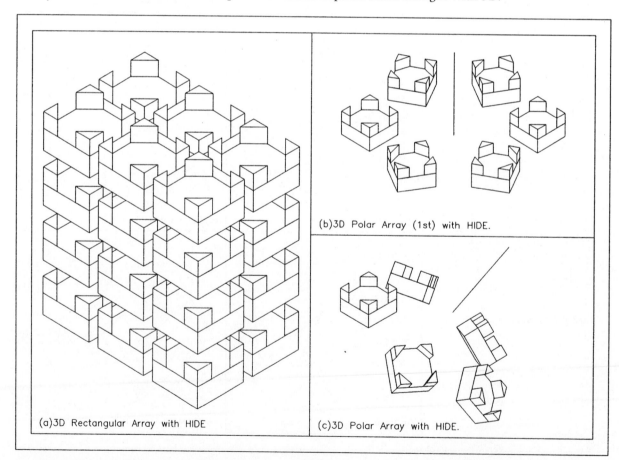

Figure 19.4 3DGEOM model – 3D array.

Three-dimensional geometry 129

1 Open the 3DGEOM model and refer to Fig. 19.4.

2 Ensure UCS BASE is current.

3 From the menu bar select **Construct**
  3D Array
  Rectangular
  prompt Select objects
  respond window the model then right-click
  prompt Number of rows (- - -)<1> and enter **2** <R>
  prompt Number of columns (| | |)<1> and enter **3** <R>
  prompt Number of levels (...)<1> and enter **4** <R>
  prompt Distance between rows and enter **120** <R>
  prompt Distance between columns and enter **120** <R>
  prompt Distance between levels and enter **100** <R>

4 Now Zoom–All. The HIDE command is useful at this stage.

5 The model is displayed in a 2 × 3 × 4 rectangular array – fig. (a).

6 Undo the rectangular array effect.

7 Draw a line from 200,200,0 to 200,200,250 then zoom–all.

8 From the Modify toolbar select the 3D Polar Array icon from the Copy flyout and:
  prompt Select objects
  respond window the model then right-click
  prompt Number of items and enter **6** <R>
  prompt Angle to fill<360> and right-click
  prompt Rotate objects ... and right-click
  prompt Center point of array
  respond **ENDpoint and pick lower end of line**
  prompt Second point on axis of rotation
  respond **ENDpoint and pick top end of line**

9 The model is array for six items about the line and is displayed in fig. (b) with hide.

10 Undo the polar array command and erase the line. Draw another line from 200,200,0 to @**100,100,200**.

11 From the menu bar select **Construct**
  3D Array
  Polar
  prompt Select objects
  respond window the model then right-click
  prompt Number of items and enter **5** <R>
  prompt Angle to fill and right-click
  prompt Rotate objects and right-click
  prompt Centre point of array
  respond **ENDpoint and pick lower end of line**
  prompt Second point on axis of rotation
  respond **ENDpoint and pick upper end of line**

12 Zoom–all the HIDE.

13 The resultant array is displayed in fig. (c) with hide.

14 This completes the 3D Array exercise. Save?

Align

The align command can be used with 2D or 3D entities and will be demonstrated using 3D objects so:

1 Open your STD3D standard sheet with layer MODEL current.

2 Using the Surfaces toolbar create the following objects as Fig. 19.5(a):

	Box	Wedge
corner	50,50,0	210,30,0
length	100	100
width	80	80
height	50	50
rotation	0	0
colour	red	green

3 The sloped surface of the wedge will be aligned onto the top of the box.

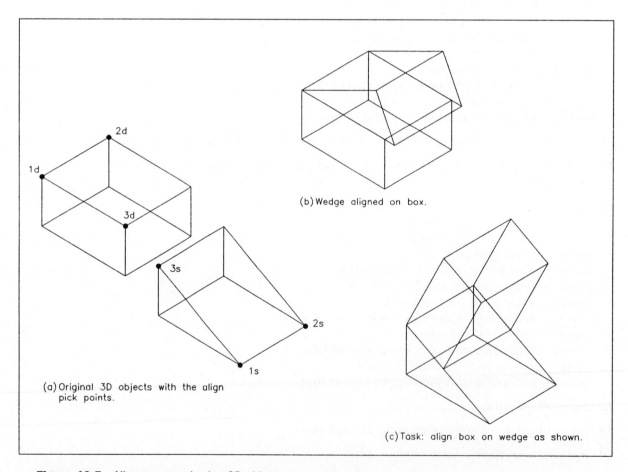

Figure 19.5 Align command using 3D objects.

4 From the menu bar select **Modify**
Align

prompt	Select objects
respond	pick the green wedge then right-click
prompt	1st source point
respond	INTersection and pick pt1s
prompt	1st destination point
respond	INTersection and pick pt1d
prompt	2nd source point
respond	INTersection and pick pt2s
prompt	2nd destination point
respond	INTersection and pick pt2d
prompt	3rd source point
respond	INTersection and pick pt3s
prompt	3rd destination point
respond	INTersection and pick pt3d

5 The green wedge is aligned on the top surface of the box – fig. (b).

6 Note the order of selecting the source/destination points.

7 Undo the alignment and then align the box onto the sloped surface of the wedge as fig. (c).

8 This completes the align exercise.

Summary

1 The three 3D rotate, mirror and align commands are specific to 3D models and are useful.
2 3D Rotate allows rotation about the X, Y and Z axes.
3 3D Mirror allows models to be 'turned' through 180°.
4 The 3D array command is the same as the 3D command but has 'levels' in the Z-direction.
5 3D models can be aligned with each other.

Chapter **20**

Dynamic viewing

Dynamic viewing is a powerful (yet underused) addition to 3D modelling as it allows models to be viewed from a perspective viewpoint. The command also allows objects to be 'cut-away' enabling the user to 'see inside' the model.

Dynamic viewing has its own terminology which is obvious when you know what is happening, but can be confusing to new users of the command.

The basic concept is that the user has a **CAMERA** which is positioned at a certain **DISTANCE** from the model – the **TARGET**. The user is looking through the camera lens at the target and can **ZOOM** in/out as required. The viewing direction is from the camera lens to a particular point on the target. The camera can be moved relative to the target and both camera and target can be twisted relative to each other. Two other concepts which the user will encounter with the command are the **slider bar** and the **perspective icon**. The slider bar allows the user to 'scale' the variable which is current, while the perspective icon is displayed when the perspective view is on.

These dynamic view concepts are illustrated in Fig. 20.1(A) with:

fig. (a): the basic terminology
fig. (b): the slide bar
fig. (c): the perspective icon.
The dynamic view command has twelve options, these being:

CAmera; TArget; Distance; POints; PAn; Zoom; TWist; CLip; Hide; Off; Undo; eXit

The option required is activated by entering the CAPITAL letters at the command line, e.g. CA for camera, TW for twist, etc. We will investigate these options with two examples:
• AutoCAD's house
• a previously saved drawing.

Example 1: AutoCAD's house

AutoCAD has a 'drawing' – actually a type of block – which will be used to demonstrate the dynamic view command and it's various options, so:

1 Begin a new drawing, accepting the prototype default and refer to Fig. 20.1(B)

2 From the menu bar select **View**
                     3D Dynamic View
prompt   Select objects
respond   right-click as nothing on screen
a) prompt   CAmera/TArget ...
    and   some cyan, red and black lines appear
    enter   **Z** <R> – the zoom option
    prompt   Slider bar with scale
    and   Adjust zoom scale factor<1>
    enter   **0.5** <R>
    result   full plan view of house – fig. (a)

b) prompt   CAmera/TArget ...
    enter   **CA** <R> – the camera option
    prompt   3D ghost image of house
    and
    enter   Toggle angle in/Enter angle from XY plane
    30 <R>
    prompt   Toggle angle from/Enter angle in *XY* plane from *X* axis
    enter   **30** <R>
    result   3D view of house – fig. (b)
c) prompt   CAmera/TArget ...
    enter   **H** <R> – the hide option
    result   house is displayed with hide – fig. (c)
d) prompt   CAmera/TArget ...
    enter   **CL** <R> – the clip option
    prompt   Back/Front/<Off>
    enter   **F** <R> – the front option
    prompt   Eye/ON/OFF/<Distance from target><1>
    enter   **40** <R>
    prompt   CAmera/TArget ...
    enter   **H** <R>
    result   house is displayed 'clipped away' at the front as is shown in fig. (d).
e) enter   **U** <R> – undoes the hide option
    enter   **U** <R> – undoes the clip option
    enter   **U** <R> – undoes the hide option
    and   **leave house as it is and read before proceeding.**

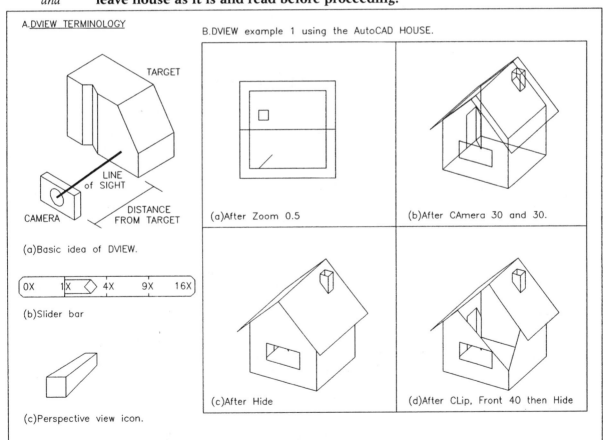

Figure 20.1 DVIEW terminology and AutoCAD's house.

134 Modelling with AutoCAD

Explanation of the command

Dynamic view is an interactive command and the various options can be used one after the other. The undo (U) option will undo the last option performed, and can be used repeatedly until all the options used have been undone. We have used some of the options to demonstrate the command, these options being zoom, camera, clip, hide and undo. The hide option is very useful, as it allows the model to be displayed when other options have been used and removes the 'ambiguity' from the model. The command can be used with wire-frame, surface and solid models. We have left the 'house' displayed on the screen, with the prompt line showing the 12 options and **you** will now investigate these various options.

CAmera

This option is used to direct the camera at the object target, and the camera can be tilted relative to two angles:
prompt 1: the angle in the XY-plane – between $-90°$ and $90°$
prompt 2: the angle form the XY-plane – between $-180°$ and $180°$.

The angles can be:
a) entered directly from the keyboard
b) toggled in using the ghost image as a guide.

Figure 20.2(A) demonstrates the option (with HIDE) for the following angle value entries:

	angle in XY-plane	angle from XY-plane
a)	35	35
b)	35	−35
c)	−35	35
d)	−35	−35

This option can be considered similar to the **VPOINT ROTATE** command.

TArget

Allows the target (the model) to be tilted relative to the position of the camera. The two angle prompts are the same as the camera option, and Fig. 20.2(B) displays the following entries:

	angle in XY-plane	angle from XY-plane
a)	35	35
b)	35	−35
c)	−35	35
d)	−35	−35

This option can be used to produce the same effect as the camera option, remembering that the camera and target are be tilted in the 'opposite sense' to each other.

TWist

This option allows the plane on which the target is 'resting' to be twisted through an entered angle. The user is prompted to enter to **new view twist**. Figure 20.3(A) displays four different twist angles for the house, the CAmera being set at 30, 30. These twist angles are:

a) 35 b) −35 c) 180 d) −90

This is a very useful option, as it allows models to be 'flipped' over by $180°$.

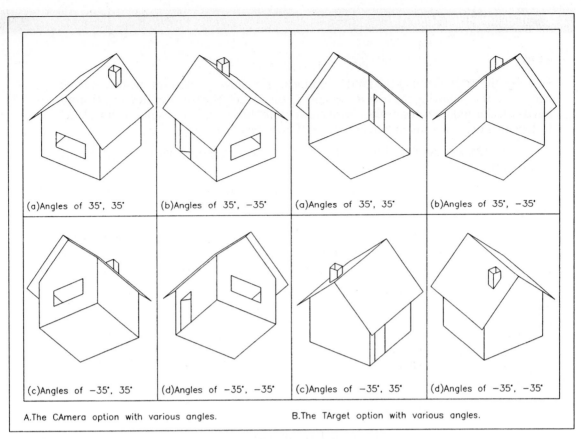

Figure 20.2 The dynamic view CAmera and TArget options.

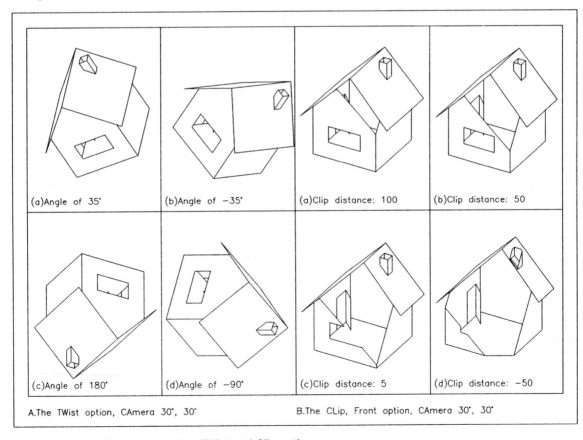

Figure 20.3 The dynamic view TWist and Clip options.

CLip

Probably the most useful of the DVIEW options, as it allows sections of the model to be 'cut-away' thus allowing the user to 'see inside' the model. The user decides on a (F)ront or (B)ack clip and then enters the clip distance from the target. The resultant clip 'plane' is dependent on the camera 'orientation' and is perpendicular to it.

Figure 20.3(B) displays four different front clip views of the house with the CAmera set at 30,30, these being:

a) distance 100   *b*) distance 50   *c*) distance 0   *d*) distance −50

POints

The points option allows the model to be viewed from a specific 'stand point', the user looking at a point on the target. Two coordinate inputs are required – the target point and the camera point. Figure 20.4 displays eight different POint displays of the house, these being:

	target point	camera point
a)	0,0,0	30,0,0
b)	0,0,0	0,30,0
c)	0,0,0	0,0,30
d)	0,0,0	30,30,0
e)	0,0,0	30,0,30
f)	0,0,0	30,30,30
g)	10,20,30	0,0,0
h)	0,0,0	10,20,30

This option is very similar to the VPOINT vector command.

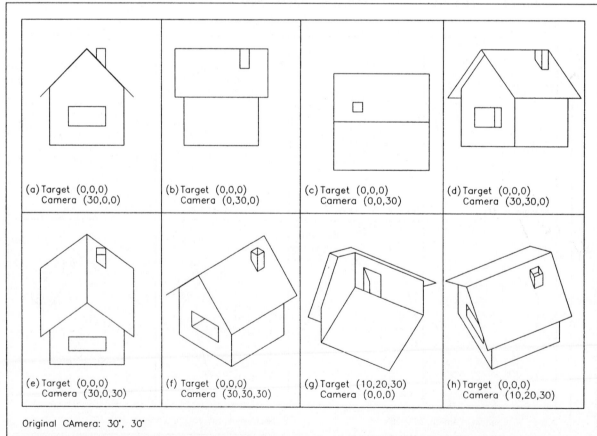

Figure 20.4 The dynamic view POints option.

Distance

Alters the distance between the camera and the target, the user entering/toggling the new distance. This option introduces **true perspective** to the model. Figure 20.5(A) displays the house with the following distance values:

a) 1000 *b)* 1500 *c)* 2500 *d)* 5000

Zoom

Does what it says – 'zooms the model'. The zoom scale factor can be entered from the keyboard, or toggled with the slider bar. The effect is displayed in Fig. 20.5(B) with zoom factors of:

a) 1 *b)* 0.75 *c)* 0.5 *d)* 0.25

PAn

An option which is the same as the ordinary PAN command, the model being 'panned' between two points.

Hide

Will display the model with a hide effect and thus removes any ambiguity from the users mind.

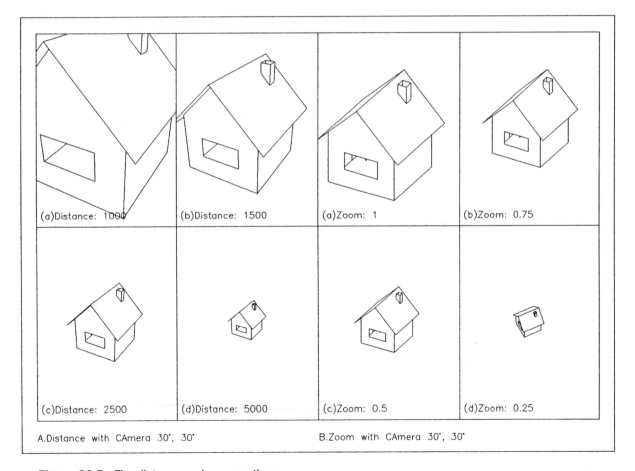

Figure 20.5 The distance and zoom options.

Undo

Entering **U** <R> will undo the last option and can be used repetitively until all the options entered have been undone.

eXit

The **X** <R> option will end the DVIEW command sequence and a blank screen will be obtained. This is because the 'house' is not a 'real drawing', but an interactive aid to using the command. When 'real models' are used with the command, the model orientation will be dependent on the DVIEW options used.

Note

1 The various options can be used continuously and repetitively and Fig. 20.6 displays some continuous sequences.
2 The various displays in my house drawings may be slightly different from your display. This is because I have had to 'adjust' the house to fit into the various viewports.
3 Before leaving the 'house', try some different options to become more familiar with the command.

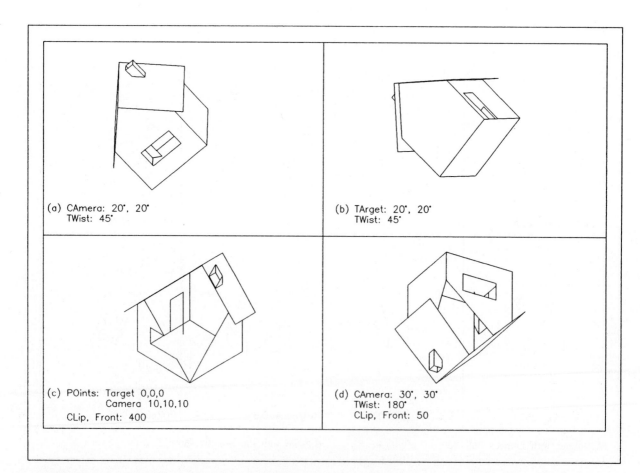

Figure 20.6 Continual combined DVIEW options.

Example 2: a saved drawing

In this example we will use a previously created model to investigate the DVIEW command, so:

1. Open model FLOWBED which was created in Chapter 13 and refer to Fig. 20.7. The original model is displayed in fig. (a).

2. Set a four viewport configuration and centre the model in each viewport – easy?

3. With the top left viewport current, enter **DVIEW** <R> at the command line and:
   - *prompt*   Select objects
   - *respond*   **window the model** then right-click
   - *prompt*   CAmera/TArget ...
   - *enter*   **CA** <R>
   - *then*   **35** <R> and **20** <R> as the angles
   - *prompt*   CAmera/TArget ...
   - *enter*   **H** <R> – fig. (b)
   - *then*   **X** <R> – to exit the command.

4. With the top right viewport active, select from the menu bar **View–3D Dynamic View** and:
   - *prompt*   Select objects
   - *respond*   **window the model** then right-click
   - *prompt*   CAmera/TArget ...
   - *enter*   **TW** <R> then **90** <R> as the new twist
   - *then*   **H** <R> – fig. (c)
   - *then*   **X** <R>

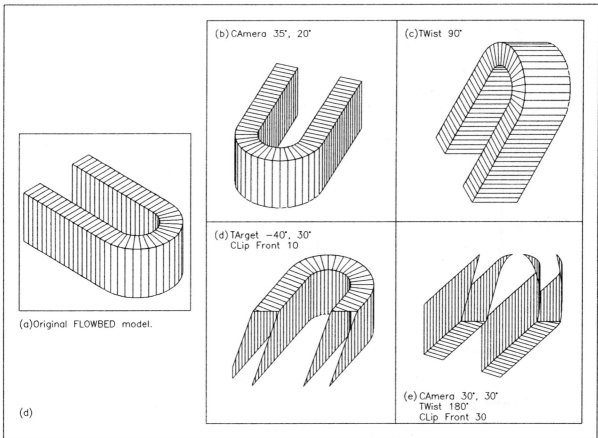

Figure 20.7 Dynamic view FLOWBED (with HIDE).

140 *Modelling with AutoCAD*

5 Use the dynamic view command in the lower viewports, window the model then enter the following options:
a) lower left viewport
   TA: with angles of –40 and 30
   CL: Front with distance of 10
   H – fig. (d)
   X
b) lower right viewport
   CA: with angles of 30 and 30
   TW: with new twist of 180
   CL: Front with distance of 30
   H – fig. (e)
   X

Note: you may have to alter the front clip distance values to obtain the same display as mine. This is due to the viewports I have created being 'smaller' than your configurations.

6 Now SHADE each viewport to 'see' the model orientation and note the cut-away effect
i) cut away from the back
ii) inverted effect.

7 Save your screen configuration if required.

Summary

1 The dynamic view is viewport specific, i.e. it only affects the model in the active viewport.
2 The camera and target options are similar to the VPOINT rotate command, and both options can be used to give the same result.
3 The twist option is useful, as it allows models to be 'inverted'
4 The distance option introduces 'true perspective'
5 The clip option allows models to be 'cut-away' to 'see inside' the model.
6 The command can be used:
a) directly on models
b) interactively using the AutoCAD 'house'.
7 The command is very powerful but underused – probably because most users are not 'comfortable' with it.

Activity

It is some time since you have been asked to attempt any activity, but we will now remedy this. The activity is to investigate a model which has already been created, so:

1 Open the model **MVMOD** created in Chapter 18. This model was created from 3D Objects and had a red cuboid positioned inside a blue cuboid. This red cuboid cannot be seen with the HIDE command.

2 Set a four viewport configuration with 3D Viewpoint Presets set to:
a) top left: NW Isometric   *b*) top right: NE Isometric
c) bottom left: SW Isometric   *d*) bottom right: SE Isometric

3 In each viewport use the dynamic view command to cut away the model so that the red cuboid can be 'seen'.
Note: I only used the CLIp–Front option.

4 Hide and shade each viewport.

Chapter **21**

Viewport specific layers

When layers are used with multiple viewports they are generally **global**, i.e. what is drawn on a layer in one viewport will be displayed in the other viewports of the drawing. This is acceptable for creating models but is unacceptable for certain other concepts, e.g. when dimensioning a model. If dimensions are to be added to a multi-view drawing, then these dimensions should only be visible in the viewport which is active.

In this chapter we will investigate how to:

1 insert a 3D model into a drawing

2 create and use viewport specific layers.

Example 1: global layer dimensions

This example will use the model created in Chapter 19 – 3DGEOM.

1 Open your **A3MVPROT** standard sheet created during the paper space/model space chapter and:
  a) layer MODEL current
  b) model space with lower right (3D) viewport active
  c) refer to Fig. 21.1.

2 From the menu bar select **Draw**
                                  Insert
                                      Block ...
  prompt   Insert dialogue box
  respond   1 pick File ...
                2 pick your R13MODEL directory
                3 pick the 3DGEOM drawing file
                4 pick OK
  prompt   Insert dialogue box
  and   note the names!
  respond pick OK
  prompt   Insertion point
  enter   **0,0,0** <R>
  prompt   X scale ... and enter **1** <R>
  prompt   Y scale ... and enter **1** <R>
  prompt   Rotation ... and enter **0** <R>

3 The red/blue model will be displayed in the four viewports.

4 Set the UCS to 50,50,0 (the model base corner) and save as BASE.

5 In each viewport, zoom about a centre point of 50,50,50 with 200 magnification.

6 Remember that dimensioning is a 2D concept!

142 *Modelling with AutoCAD*

7 With the lower left viewport (top) active and UCS–BASE make layer DIMS (magenta) current.

8 Linear dimension one horizontal and one vertical line of the model.

9 These two dimensions will be displayed in all viewports, thus illustrating the global effect of layers.

10 Erase the two added dimensions, then REGENALL. Do not exit the drawing.

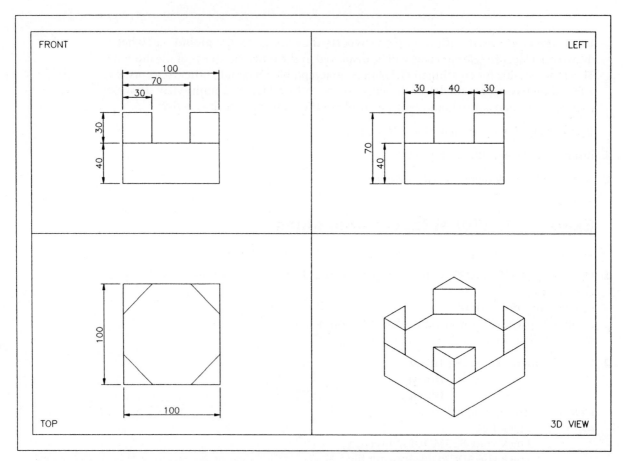

Figure 21.1 Using viewport specific layers to add dimensions to individual viewports.

Example 2: viewport specific layers

To demonstrate viewport specific layers we will make three new layers which will allow us to add dimensions to each of the top-left, top-right and bottom-left viewports.

1 Ensure the lower left viewport is active and UCS-BASE is set.

2 From the menu bar select **Data**
            Viewport Layer Controls
            New Freeze

prompt   New Viewport frozen layer name(s)
enter   **DIMTL,DIMTR,DIMBL** <R>
prompt   ?/Freeze/Thaw ...
respond right-click

3 From the menu bar select **Data–Layers** and:
prompt   Layer Control dialogue box
with   1   three new layers DIMBL, DIMTL, DIMTR
           2   all three have **C** and **N** in the State column
respond   1   pick DIMBL line
           2   pick **CurVP: Thw** – C removed?
           3   pick Current
           4   pick OK.

4 Layer DIMBL name displayed?

5 Refer to Fig. 21.1 and add the two dimensions as displayed in the lower-right viewport – and this time they should *not* be displayed in the other three viewports?

6 Make the top-left viewport active, activate the layer control dialogue box and:
a) pick DIMTL line
b) pick CurVP:Thw
c) pick Current then OK.

7 At the command line enter **UCS-X-90** to reset the UCS position.

8 Refer to Fig. 21.1 and dimension:
a) vertical using the continuous option
b) horizontally using the baseline option.

9 Dimensions should only be added to the top-left viewport.

10 Make the top-right viewport active.

11 Reset the UCS with a rotation about the *Y*-axis of −90.

12 Using the layer control dialogue box:
a) pick DIMTR
b) pick CURVP:Thw
c) pick Current then OK.

13 Now dimension:
a) vertically with baseline
b) horizontally with continuous.

14 At this stage your 3DGEOM model should be fully dimensioned as Fig. 21.1.

15 Before leaving this exercise make the lower right viewport active and using the layer control dialogue box:
a) pick DIMTL, DIMTR, DIMBL lines
b) pick CurVP:Thw
c) pick OK.

16 The lower right viewport will display all 12 dimensions.

17 If you erase any dimension it will be erased from other viewports.

18 This completes the exercise, and the drawing does not need to be saved.

Summary

1 Viewport specific layers can only be used if TILEMODE is set to 0, i.e. they are for paper space multiple viewports.

2 Viewport specific layers are described by their name – they are layers which have been created and are 'tied' to a specific viewport.

3 Viewport specific layers are used for different reasons with 3D models. Our example used dimensioning as a demonstration.

4 The specific layers are created:

a) from the menu bar with **Data–Viewport Layer Controls**

b) by entering **VPLAYER** at the command line.

5 There are two new states with viewport specific layers:

C: layer is frozen in current viewport. If no C, then the layer is **thawed (Thw)** in the current viewport.

N: layer is frozen in new viewports.

6 Viewport specific layers are a very useful and powerful tool when working with floating (paper space) multiple viewports.

Chapter **22**

Solid modelling introduction

Three-dimensional modelling with computer-aided draughting and design (CADD) can be divided into three categories:

- wire-frame modelling
- surface modelling
- solid modelling.

We have already created wire-frame and surface models and will now concentrate on how solid models are created. This chapter will summarize the three model types.

Wire-frame modelling

Wire-frame models are defined by points and lines and are the simplest possible representation of a 3D component. They may be adequate for certain 3D model representation and require less memory than the other two model types, but wire-frame models have several limitations:

1 Ambiguity: how you view the model, i.e. are you 'looking down at' or 'looking up at' the model.

2 No curved surfaces: while curves can be added to a wire-frame model, an actual curved surface cannot. Lines may be added to give a 'curved effect' but the computer does not recognize these as being part of the model.

3 No interference: as wire-frame models have no surfaces, they cannot detect interference between adjacent components. This makes them unsuitable for kinematic displays, simulations, etc.

4 No physical properties: mass, volume, centre of gravity, moments of inertia, etc. cannot be calculated.

5 No shading: as there are no surfaces, a wire-frame model cannot be shade or rendered.

6 No hidden removal: as there are no surfaces, it is not possible to display the model with 'hidden line removal'.

AutoCAD R13 allows wire-frame models to be created.

Surface modelling

A surface model is defined by points, lines and faces and a wire-frame model can be 'converted' into a surface model by adding these 'faces'. Surface models have several advantages when compared to wire-frame models some of these being:

1 Recognition and display of curved profiles.
2 Shading, rendering and hidden line removal all possible.
3 Recognition of holes.

Surface models are suited to many applications but they have some limitations which include:

1 No physical properties: other than surface area, a surface model does not allow the calculation of mass, centre of gravity, moments of inertia, etc.
2 No detail: a surface model does not allow section detail to be obtained.

Several types of surface model can be generated and these include:
a) plane and curved swept surfaces
b) swept area surfaces
c) rotated or revolved surfaces
d) splined curve surfaces
e) nets or meshes.

AutoCAD R13 allows surface models of all these types to be created.

Solid modelling

A solid model is defined by the volume the component occupies and is thus a real 3D representation of the component. Solid modelling has many advantages which include:

1 Complete physical properties of mass, volume, centre of gravity, moments of inertia, etc.

2 Dynamic properties of momentum, angular momentum, radius of gyration, etc.

3 Material properties of stress-strain.

4 Full shading, rendering and hidden detail removal.

5 Section views and true shape extraction.

6 Interference between adjacent components can be highlighted.

7 Simulation for kinematics, robotics.

Solid models are created using a **solid modeller** and there are several types of solid modeller, the two most common being:

1 Constructive solid geometry or constructive representation (CSG/CREP). The model is created from solid primitives and/or swept surfaces using Boolean operations.

2 Boundary representation (BREP). The model is represented by the edges and faces making up the surfaces, i.e. the **topology** of the component.

Comparison of model types

The three model types are displayed in:

a) Fig. 22.1: with hidden line removal
b) Fig. 22.2: as cross-sections.

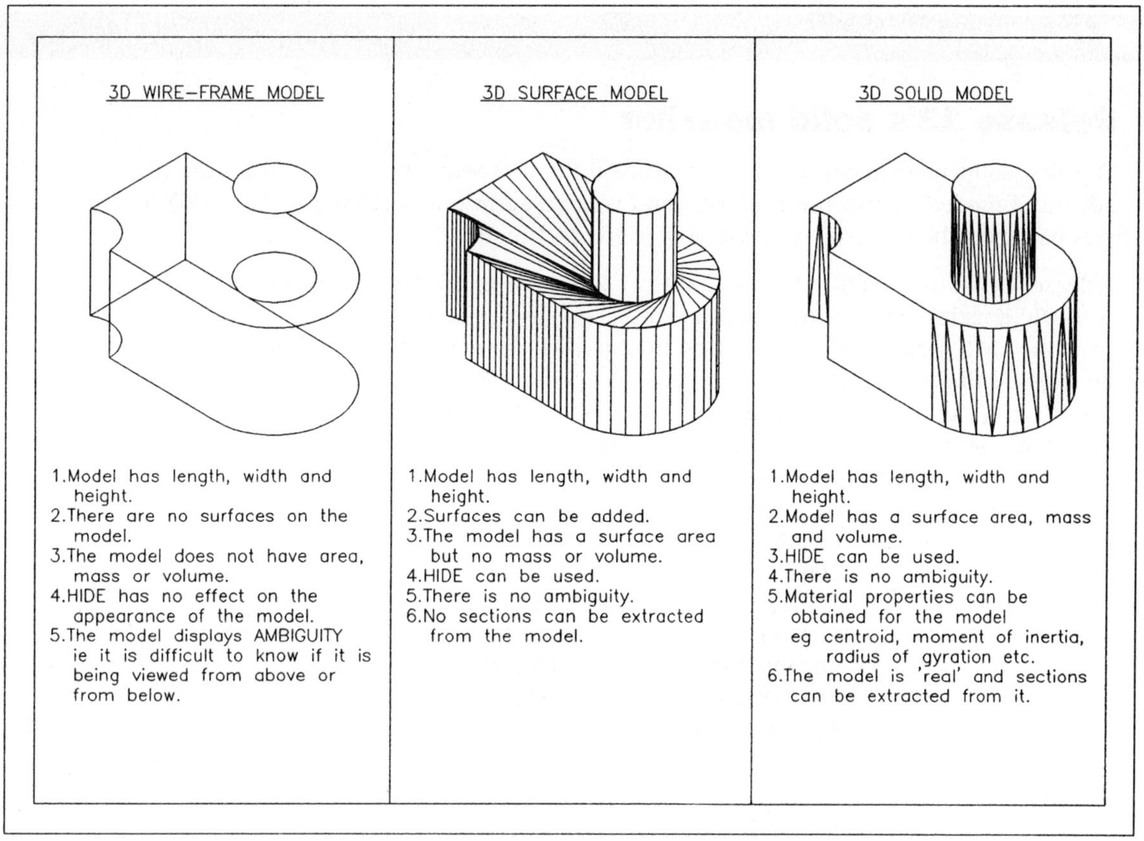

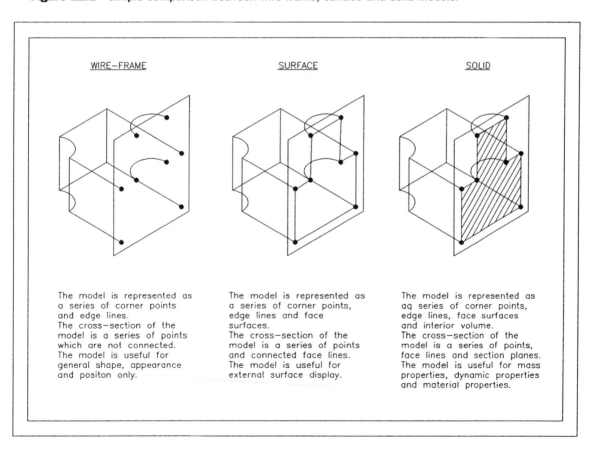

Figure 22.1 Simple comparison between wire-frame, surface and solid models.

Figure 22.2 Further comparison of model types as cross-section.

Release 13's solid modeller

The first solid modeller introduced by AutoDESK was with Release 11. This was the advanced modelling extension (AME) and was an additional package to AutoCAD. It was based on the constructive geometry (CSG) technique.

Release 13 has the solid modeller 'built in' and is based on the **ACIS** solid modeller. AutoDESK claim that this modeller has certain advantages over AME. Readers who have used the AME modeller should have no trouble with the new modeller, although it must be stressed that the R13 modeller appears to operate slightly differently from AME. It should also be pointed out that the solid modelling command structure has been completely altered and that several commands no longer exist. The following is a list of the Release 13 solid modelling commands:

		AME name	*R13 name*
a)	Creation	SOLBOX	BOX
		SOLWEDGE	WEDGE
		SOLCYL	CYLINDER
		SOLCONE	CONE
		SOLSPHERE	SPHERE
		SOLTORUS	TORUS
		SOLEXT	EXTRUDE
		SOLREV	REVOLVE
b)	Editing	SOLUNION	UNION
		SOLSUB	SUBTRACT
		SOLINT	INTERSECT
		SOLCHAM	CHAMFER
		SOLFILL	FILLET
		SOLCUT	SLICE
c)	Inquiry	SOLMASSP	MASSPROP
		SOLSECT	SECTION
		SOLINTERF	INTERFERE/STLOUT
d)	Commands	SOLAREA	with MASSPROP
	which have	SOLCHP	no longer required
	been	SOLFEAT	with regions
	withdrawn	SOLIDIFY	
		SOLIN	no longer required
		SOLLIST	no longer required
		SOLMAT	no materials, solid density is 1 Designer required for this!
		SOLMESH	automatic hide and shade
		SOLMOVE	no longer required
		SOLOUT	no longer required
		SOLPROF	not applicable
		SOLPURGE	no longer required
		SOLSEP	no longer required
		SOLUCS	not applicable
		SOLVAR	not applicable
		SOLWDENS	no longer required
		SOLWIRE	no longer required

The solid model standard sheet

As with all good draughting practice, we will create a solid model standard sheet for use with our models. This prototype drawing will:

a) be for A3-sized paper
b) have a four viewport (paper space) configuration.

We have already created a multi-view prototype drawing and we will modify this drawing and save it for all solid modelling work, so:

1 Open drawing A3MVPROT from your R13MODEL directory.

2 Screen displays four yellow viewports with FRONT, LEFT, TOP and 3DVIEW viewpoints.

3 Set a new UCS origin point with the sequence:
UCS <R>
O <R>
100,50,0 <R>

4 The icon should move to this new origin point? If not, then UCSICON-A-OR

5 Save this UCS position as BASE.

6 Set and save three other UCS positions with:

a) UCS	then	*b*) UCS	then	*c*) UCS
X		Y		Y
90		90		180
UCS		UCS		UCS
S		S		S
FRONT		RIGHT		LEFT

7 Restore UCS BASE

8 Create a text style named SOLA3 from the ROMANS text font, and accept ALL the prompt defaults.

9 Create a new dimension style named **DIMSOLID** with the following settings:

a) Geometry	Spacing: 12
	Extension: 3
	Origin offset: 3
	Arrowheads: Closed filled
	Centre mark: None
b) Format	User Defined
	Fit: Best
	Horizontal Justification: Centred
	Text: Outside Horizontal
	Vertical Justification: Above
c) Annotation	Units: Decimal to 0.00 precision
	Decimal angles
	Text: Height 6
	Gap 1.5
	Style SOLA3

10 At the command line enter **ISOLINES** <R>
prompt　New value for ISOLINES<4>
enter　　**24** <R>

150 *Modelling with AutoCAD*

11 At the command line enter **FACETRES** <R>
   prompt   New value for FACETRES<0.5>
   enter   **1** <R>

12 Ensure layer MODEL is current.

13 Activate the Draw, Modify, Object Snap and Solids toolbars.

14 Save the solid model standard sheet as **\R13MODEL\SOLA3**

15 You are now ready to start creating solid models.

Note

1 Two new system variables have been introduced in the creation of the solid model standard sheet. These are:
ISOLINES: controls the number of tessellation lines used in the visualization of curved portions of models. It is an integer value between 0 and 2047.
FACETRES: adjusts the smoothness of shaded and hidden line removal models. The value can be 0.01–10.0

2 The ISOLINES and FACETRES values may be modified with different models.

3 The origin point may also be altered as new models are created.

4 We have not zoom-centred the viewports as this will depend on the model being created.

5 Solid modelling consists of creating composites from 'primitives' and there are three types:
a) basic primitives
b) swept volume primitives
c) edge primitives.

All three types will be discussed in detail.

Chapter **23**

The basic solid primitives

Release 13 supports the six basic solid primitives: box, wedge, cylinder, cone, sphere and torus. In this chapter we will create models using each primitive, and also investigate the various options which are available. During the exercises do not just accept the coordinate values given, try and reason out why they are being used.

The BOX primitive – Fig. 23.1

1 Open your SOLA3 standard sheet with layer MODEL current and the UCS set to BASE. The lower-right (3D) viewport should be active. Ensure that the Solids toolbar is displayed.

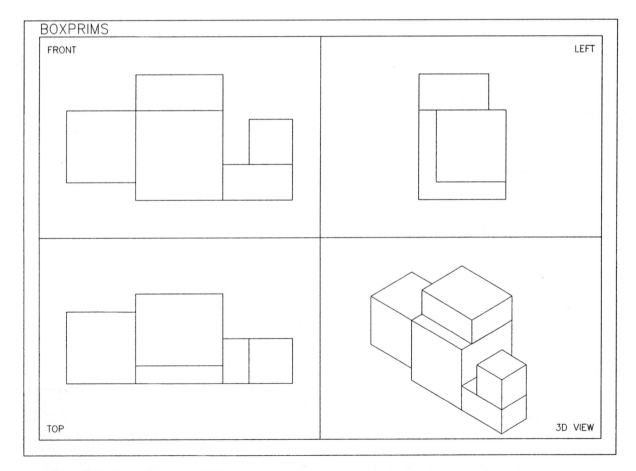

Figure 23.1 The BOX solid primitives.

152 *Modelling with AutoCAD*

2 From the menu bar select **Draw**
                            Solids
                            Box
                            Corner
   prompt   Center/<Corner of box><0,0,0>
   enter   **0,0,0** <R>
   prompt   Cube/Length/<other corner>
   enter   **C** <R> – the cube option
   prompt   Length
   enter   **100** <R>

3 A red cube in displayed in all viewports.

4 Select the corner box icon from the Solids toolbar and:
   prompt   Center/<Corner of box>
   enter   **100,0,0** <R>
   prompt   Cube/Length ...
   enter   **L** <R> – the length option
   prompt   Length and enter **80** <R>
   prompt   Width and enter **50** <R>
   prompt   Height and enter **40** <R>

5 Another red cuboid will be displayed. This cuboid is to be a different colour so at the command line enter **CHPROP** <R> and:
   prompt   Select objects
   respond **pick the last box created** and right-click
   prompt   Change what properties ...
   enter   **C** <R> – for colour
   prompt   New colour ...
   enter   **3** <R> – for green
   prompt   Change what properties ...
   respond right-click

6 At the command line enter **BOX** <R> and:
   prompt   Center/<Corner of box>
   enter   **0,0,100** <R>
   prompt   Cube/Length ...
   enter   **L** <R>
   prompt   Length and enter **–80** <R>
   prompt   Width and enter **80** <R>
   prompt   Height and enter **–80** <R>

7 Change the colour of this box to yellow.

8 Repeat the BOX command and:
   prompt   Center ...
   enter   **100,100,100**
   prompt   Cube/Length/<other corner>
   enter   **@–100,–80,40** <R> – the other corner option colour blue

9 Finally select the center box icon from the Solids toolbar and:
   prompt   Centre of box
   enter   **155,25,65** <R>
   prompt   Cube/Length ...
   enter   **C** <R>
   prompt   Length and enter **25** <R> colour magenta

10 Now centre the model in each viewport:
  a) about the centre point 50,50,70
  b) at 225 magnification (275 in 3D viewport).

11 Question: why a centre point of 50,50,70?

12 From the menu bar select **Tools-Hide** in each viewport.

13 At the command line enter **SHADE** <R> in each viewport and you should have nice coloured blocks displayed.

14 Save the model creation as BOXPRIMS.

The WEDGE Primitive – Fig. 23.2

1 Open your SOLA3 standard sheet, MODEL layer, lower right viewport active, UCS BASE current and Solids toolbar.

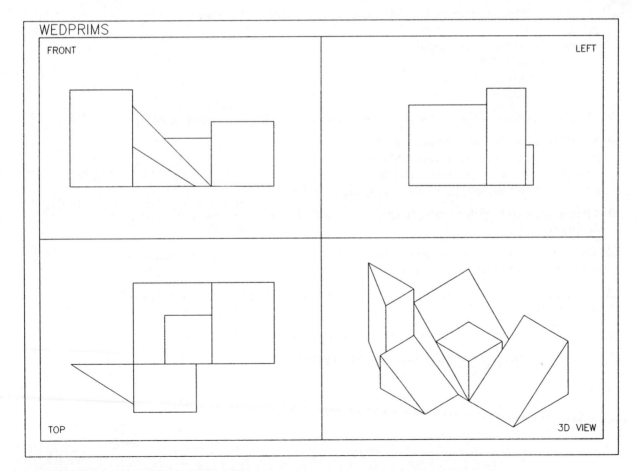

Figure 23.2 The WEDGE solid primitive.

154 *Modelling with AutoCAD*

2 From the menu bar select **Draw**
   Solids
   Wedge
   Corner
   prompt   Center/<Corner of wedge><0,0,0>
   enter   **0,0,0** <R>
   prompt   Cube/Length/<other corner>
   enter   **C** <R> – the cube option
   prompt   Length and enter **100** <R>

3 A red wedge is displayed with its base corner at 0,0,0.

4 Select the Wedge Corner icon from the Solids toolbar and:
   prompt   Center/<Corner of wedge><0,0,0>
   enter   **0,0,0** <R>
   prompt   Cube/Length ...
   enter   **L** <R> – the length option
   prompt   Length and enter **80** <R>
   prompt   Width and enter **–60** <R>
   prompt   Height and enter **50** <R>

5 Change the colour of this wedge to blue.

6 Enter **WEDGE** <R> at the command line and:
   prompt   Center/<Corner ...
   respond   **ENDpoint and pick right-most vertex of the red wedge**
   prompt   Cube ...
   respond   select the Length option with:
               Length: 100
               Width: 80
               Height: 80

7 Change the colour of this wedge to yellow, then rotate it about its corner point by –90.

8 Select the corner wedge icon again and:
   prompt   Corner point and enter **100,0,0** <R>
   prompt   Cube/Length/<other corner>
   enter   **@–60,60,–60** <R> – the other corner option.

9 Change the colour of this wedge to green, then MOVE it from any one of its vertices by **@0,0,60**.

10 The final wedge is to be created with its corner point at 0,0,0 and with a length of –80; width 120 and height 50. Its colour is to be magenta.

11 Use the 3D ROTATE command to rotate the magenta wedge about the the *X*-axis point 0,0,0 by 90°.

12 Centre the four viewports about the point 60,30,60 with 250 magnification.

13 Hide and shade then save your model as WEDPRIMS.

The CYLINDER primitive – Fig. 23.3

1. Open the SOLA3 standard sheet as before.
2. From the menu bar select **Draw**
   Solids
   Cylinder
   Center
   prompt Elliptical/<center point>
   enter **0,0,0** <R>
   prompt Diameter/<Radius>
   enter **25** <R>
   prompt Centre of other end/<Height>
   enter **80** <R>

3. Select from the menu bar **Draw–Solids–Cylinder–Elliptical** and:
   prompt Center/<Axis endpoint>
   enter **C** <R> – the centre option
   prompt Center of ellipse and enter **40,0,0** <R>
   prompt Axis endpoint and enter **@–15,0,0** <R>
   prompt Other axis endpoint and enter **@25,0,0** <R>
   prompt Center of other end/<Height>
   enter **60** <R>

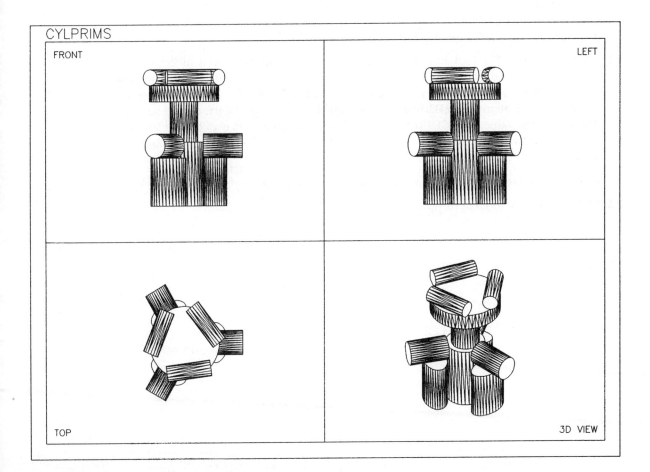

Figure 23.3 The CYLINDER solid primitive.

156 *Modelling with AutoCAD*

4 Change the colour of this cylinder to yellow, then polar array it:
   a) about the centre point 0,0
   b) for three items with rotation.

5 Centre each viewport about the point **0,0,80** with **0.5XP**.

6 Using the Cylinder Centre icon from the Solids toolbar create two cylinders from the following:

Centre pt	rad	ht	colour
0,0,80	18	50	green
0,0,130	45	20	blue

7 At the command line enter **CYLINDER** <R> and:
   prompt   Elliptical ... and enter **25,0,75** <R>
   prompt   Diameter ... and enter **15** <R>
   prompt   Center of other end/<Height>
   enter   **C** <R> – the other end option
   prompt   Center of other end
   enter   **@50<0<0** <R>

8 Change this cylinder's colour to cyan then polar array it about the point 0,0 for three items with rotation.

9 Finally create a magenta cylinder:
   a) centre point 45,0,160
   b) radius 10
   c) other end at 10,45,160
   d) polar array about 0,0 for three items.

10 Hide the model – Fig. 23.3 and note the 'triangular facets' on the curved surfaces. These facets are controlled by the FACETRES system variable which was set to 1 in our standard sheet. At the command line enter FACETRES and then enter a value of 5. Use the REGENALL command then HIDE again and note the appearance of the curved surfaces. The higher the value of FACETRES then the 'better the appearance' of curved surfaces, but the model takes longer to be displayed, i.e. it is a trade-off between appearance and speed.

   At our level of modelling, a FACETRES (facet resolution) value of 1 or 2 is suitable.

11 Shade then save as CYLPRIMS.

The CONE Primitive – Fig. 23.4

1 Open SOLA3 as before with the usual layer, etc.

2 From the menu bar select **Draw–Solids–Cone–Center**.
 prompt Elliptical/<Center point>
 enter **0,0,0** <R>
 prompt Diameter/<radius> and enter **50** <R>
 prompt Apex/<Height> and enter **60** <R>

3 At the command line enter **CONE** <R> and:
 prompt Elliptical/<center point> and enter **0,0,0** <R>
 prompt Diameter/<Radius>
 enter **D** <R> – the diameter option
 prompt Diameter and enter **160** <R>
 prompt Apex/<Height> and enter **–80** <R>

4 Change the colour of this inverted cone to green.

5 In each viewport zoom centre about 0,0,0 at 0.5XP.

Figure 23.4 The CONE solid primitive.

158 *Modelling with AutoCAD*

6 Select the Elliptical Cone icon from the Solids toolbar and:
   prompt Center/<axis endpoint>
   enter **C** <R> – the centre option
   prompt Center of ellipse and enter **70,0,0** <R>
   prompt Axis endpoint and enter **@10,0,0** <R>
   prompt Other axis distance and enter **@0,25,0** <R>
   prompt Apex/<Height> and enter **40** <R>

7 Change the colour of this elliptical cone to yellow then polar array it about the point 0,0 for seven items with rotation.

8 At the command line enter **CONE** <R> and:
   prompt Elliptical ...
   enter **80,0,0** <R> as the centre point
   prompt Diameter ... and enter **30** <R>
   prompt Apex/<Height>
   enter **A** <R> – the apex option
   prompt Apex
   enter **@50,0,0** <R>

9 Change the colour of the cone to blue, then 3D ROTATE it:
   a) about the *Y*-axis
   b) about the point 80,0,0
   c) with 45° rotation.

10 Polar array the blue cone about 0,0 for five items with rotation.

11 Create the final cone, colour magenta with:
   a) centre point 0,0,80
   b) diameter 40
   c) apex at @0,–100,0.

12 Hide and shade each viewport then REGENALL.

13 Save the drawing as CONPRIMS.

The SPHERE primitive – Fig. 23.5

1 Open the SOLA3 standard sheet as before.

2 From the menu bar select **Draw**
                              Solids
                              Sphere
   prompt Center of sphere
   enter   **0,0,0** <R>
   prompt Diameter/<Radius> of sphere
   enter   **60** <R>

3 Centre the four viewports about the point 0,0,20 at 0.6XP.

4 At the command line enter **SPHERE** <R> and:
   prompt Center of sphere
   enter   **80,0,0** <R>
   prompt Diameter/<Radius>
   enter   **D** <R> – the diameter option
   prompt Diameter and enter **40** <R>.

5 Change the colour of this sphere to yellow, and polar array it about the point 0,0 for five items with rotation.

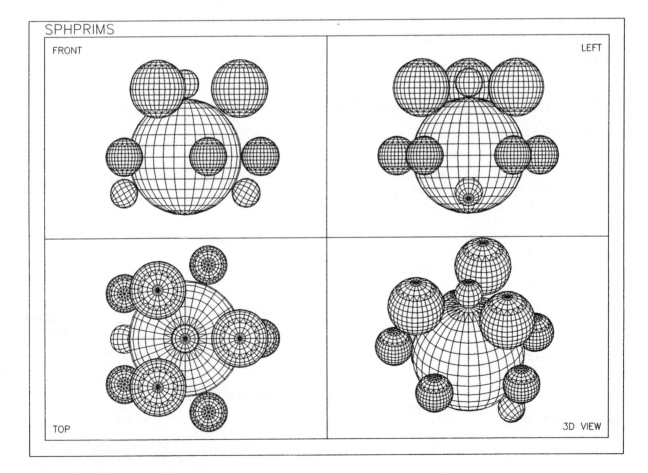

Figure 23.5 The SPHERE solid primitive.

160 *Modelling with AutoCAD*

6 Select the Sphere icon from the Solids toolbar and:
 prompt Center of sphere
 enter **0,0,75** <R>
 prompt Diameter ... and enter **15** <R>.

7 Change the colour of this sphere to green.

8 Restore the UCS FRONT.

9 Polar array the green sphere about the point 0,0 for three items with rotation.

10 Restore UCS BASE.

11 Create a sphere with:
 a) centre at 58,0,70
 b) radius 30
 c) colour blue
 d) polar arrayed about 0,0 for three items.

12 Hide, shade, regen, save as SPHPRIMS.

13 *Task*: Before leaving this exercise, enter **FACETRES** <R> at the command line and:
 prompt New value ...<1>
 enter **5** <R>

 REGENALL, then HIDE and note the appearance of the spheres. They should now be 'better defined' due to the value of the FACETRES system variable.

14 Our FACETRES original value of 1 is suitable for creating the model, but a higher value is more suited for the hide effect.

15 Now proceed to the next exercise.

The TORUS primitive – Fig. 23.6

1 Open SOLA3, etc.

2 From the menu bar select **Draw**
 Solids
 Torus
 prompt Center of torus
 enter **0,0,0** <R>
 prompt Diameter/<Radius> of torus
 enter **50** <R>
 prompt Diameter/<radius> of tube
 enter **20** <R>

3 Create another torus by selecting the Torus icon from the Solids toolbar with:
 a) centre point: 110,0,0
 b) torus radius: 30
 c) tube radius: 10
 d) colour: green.

4 Polar array the green torus about the point 0,0 for five items with rotation.

5 Centre all viewports about the point 0,0,80 at 0.5XP.

The basic solid primitives **161**

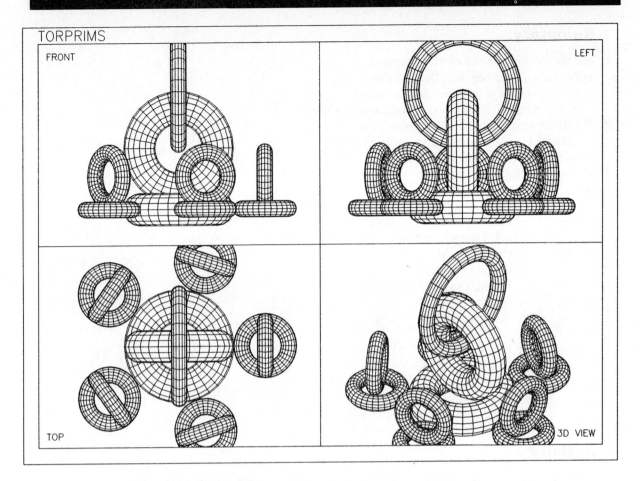

Figure 23.6 The TORUS solid primitive.

6 Create another three torus primitives using the following:

	torus1	torus2	torus3
UCS	FRONT	RIGHT	RIGHT
centre	0,75,0	0,145,0	0,40,110
torus radius	50	65	30
tube radius	20	10	10
colour	blue	cyan	yellow

7 Restore the UCS BASE then polar array the yellow torus about the point 0,0 for five items with rotation.

8 Hide, shade, regenall, save as TORPRIMS.

9 *Task*: investigate the FACETRES system variable.

162 *Modelling with AutoCAD*

Summary

1 The six solid primitives can be activated:
a) from the menu bar with Draw–Solids
b) by icon selection from the Solids toolbar
c) by entering the solid name at the command line.
2 The corner/centre points of the solids can be:
a) entered as coordinates at the command line
b) referenced to existing entities.
3 All six primitives have options:

box –   a) corner; centre
             b) cube; length, width, height; other corner

wedge –   a) corner; centre
             b) cube; length, width, height; other corner

cylinder – a) circular; elliptical
             b) diameter; radius
             c) height; centre of other end

cone –   a) circular; elliptical
             b) diameter; radius
             c) height; apex

sphere – a) centre only
             b) diameter; radius

torus –   a) centre only
             b) torus: diameter; radius
             c) tube: diameter; radius.

Activity

Refer to Tutorial 20 and create a 'solid primitive layout' using one of each of the six primitives. The following information is given for the primitives, but the **layout idea is to be your own**.

Box		Wedge		Cylinder	
Length: 100		Length: 100		Radius: 40	
Width: 100		Width: 100		Height: 40	
Height: 80		Height: 80		Colour: green	
Colour: red		Colour: yellow			

Cone		Sphere		Torus	
Radius: 50		Radius: 50		Torus rad: 80	
Height: 100		Colour: magenta		Tube rad: 20	
Colour: blue		Colour: cyan			

Chapter 24

The swept solid primitives

Solids can be generated by extruding or revolving 'shapes', and in this chapter we will use several different exercises to demonstrate how quite complex solids can be obtained from relatively simple shapes.

Extruded solids

Solids can be created by extruding objects (closed polylines, polygons, circles, ellipses, closed splines, regions) along a path or by a specified height and taper angle.

Refer to Fig. 24.1 which details in 2D the basic shapes and sizes which will be used for the extrusion exercises. The actual size of each model is really unimportant but I would suggest that you try to make the 'overall' size the same as mine as this will assist with the zoom–centre effect.

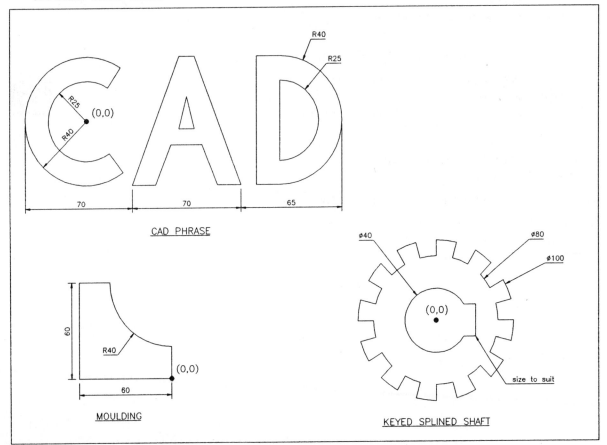

Figure 24.1 Models for use with the extrusion swept primitive exercise.

Extruded model 1: the phrase CAD

1. Open your SOLA3 standard sheet with layer MODEL current.

2. Change the viewpoint in the top right viewport to RIGHT, then make the top left viewport active. Restore UCS FRONT.

3. Using the sizes in Fig. 24.1 as a guide, create the phrase CAD from lines and circles, then trim as required. Make the centre point of circles of letter C at 0,0.

4. Using **Modify–Edit Polyline** change the two lines and two trimmed circles of the letter C into a single polyline using the (J)oin option.

5. Repeat the Edit Polyline command for the letters A and D, the command having to be used twice with each letter.

    Note: we want the extruded shape as a polyline, hence the need to 'convert' the letters from lines/arcs to polylines.

6. In each viewport zoom-centre about the point 60,0,40 at 0.75XP.

7. From the menu bar select **Draw**
    Solids
    Extrude
    prompt   Select objects
    respond **pick the C polyline** then right-click
    prompt   Path/<Height of Extrusion>
    enter   **80** <R>
    prompt   Extrusion taper angle<0>
    enter   **0** <R>

8. The letter C is extruded 80 in the positive Z-direction.

9. Select the Extrude icon from the Solids toolbar and:
    prompt   Select objects
    respond **pick the two A polylines** then right-click
    prompt   Path/<Height of Extrusion>
    enter   **50** <R>
    prompt   Extrusion taper angle<0>
    enter   **3** <R>

10. Change the colour of the letter A extruded solid to blue.

11. At the command line enter **EXTRUDE** <R> and:
    prompt   Select objects
    respond **pick the two D polylines** the right-click
    prompt   Path/<Height of ... and enter **20** <R>
    prompt   Extrusion taper ... and enter **0** <R>

12. The extruded letter D is to be green.

13. Hide each viewport to give Fig. 24.2. Shade if required.

14. Save this model as SW1PRIMS.

The swept solid primitives **165**

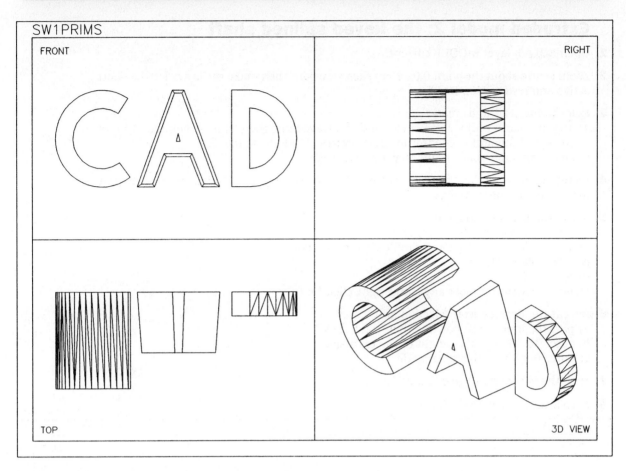

Figure 24.2 The CAD phrase as a swept primitive.

Extruded model 2: the keyed splined shaft

1. Open SOLA3, layer MODEL current.

2. Zoom centre about the point 0,0,–20 in each viewport, then make the lower left viewport active and restore UCS BASE.

3. Refer to Fig. 24.1 and create:
   a) an outer tooth profile from circles and an arrayed line then trim as required. A bit of work and thought needed? The circle centres should be at 0,0
   b) create an inner shaft profile to your own size.

4. Using Modify–Edit Polyline, convert the outer tooth profile into a single polyline. Repeat for the inner profile. More work?

5. Select the Extrude icon and:
   prompt   Select Objects
   respond pick the outer tooth profile then right-click
   prompt   Path/<Height ...
   enter   **–70** <R>
   prompt   Extrusion taper angle and right-click for 0.

6. Repeat the Extrude icon selection and:
   a) pick the inner shaft polyline then right-click
   b) enter **30** <R> as the height of the extrusion
   c) enter **0** <R> as the taper angle.

7. Hide the viewports to give Fig. 24.3.

8. Save if required as SW2PRIMS.

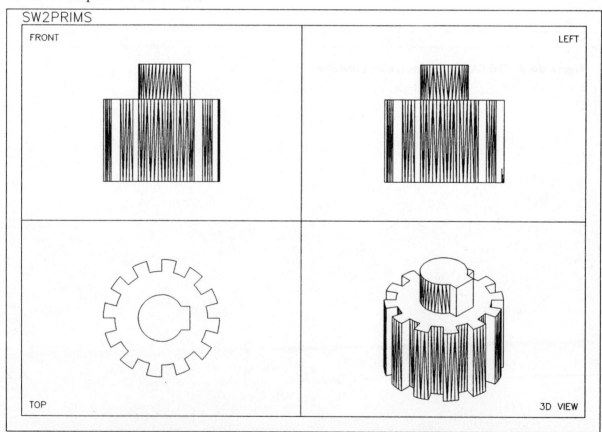

Figure 24.3 The keyed splined shaft as a SWEPT solid.

Extruded model 3: the moulding

1 Open SOLA3, etc.

2 Zoom centre about 100,120,30 at 0.4XP in each viewport.

3 With the lower-right viewport active, restore UCS BASE then draw a polyline:
From   0,0
To   @0,100
To   Arc endpoint @100,100
To   Arc endpoint @100,–50
To   right-click.

4 Change the colour of this polyline to blue.

5 Make the top-left viewport active and restore UCS FRONT.

6 Refer to Fig. 24.1 and create the moulding shape as a single polyline, the lower-right corner being at the point 0,0.

7 Select the Extrude icon and:
prompt   Select objects
respond   pick the red moulding outline then right-click
prompt   Path/<Height of ...
enter   **P** <R> – the path option
prompt   Select path
respond   pick the blue polyline.

8 The red moulding shape is extruded along the blue path.

9 Hide each viewport – Fig. 24.4.

10 Save if required.

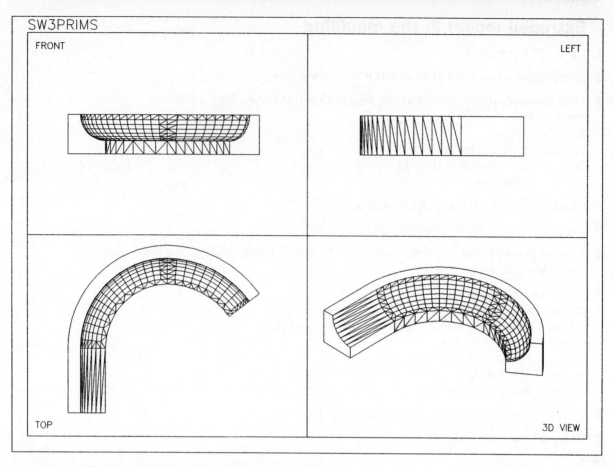

Figure 24.4 The moulding extruded along a path.

Extruded model 4: a twisted pipe

1 Open the SOLA3 standard sheet.

2 Make two new layers, P1 colour blue and P2 colour green.

3 With the lower right viewport active, layer MODEL current, restore the UCS FRONT.

4 Draw a polyline: From 0,0
  To   @0,100
  To   Arc endpoint @100,100
  To   right-click

5 *a*) Restore UCS BASE
  b) make layer P1 (blue) current
  c) draw a circle, centre at 0,0 with radius 30.

6 Select the Extrude icon and:
  a) pick the blue circle as the objects
  b) enter P <R> for the path option
  c) pick the red polyline as the path.

7 The blue circle is extruded along the red path.

8 *a*) Make layer MODEL current
  b) ensure UCS BASE is current.

9 Draw a polyline From   100,0
                 To   Arc endpoint @100,100
                 To   Line endpoint @0,100
                 To   right-click

10 Move this polyline:
   a) from 100,0
   b) by @0,0,200.

11 *a*) Restore UCS RIGHT
   b) make layer P2 (green) current
   c) draw a circle, centre at 0,200,100 with radius 30.

12 Using the Extrude icon:
   a) enter L <R><R> to select the green circle
   b) enter P <R> for the path option
   c) pick the new red polyline as the path.

13 The green circle is extruded along the red path.

14 In each viewport, zoom centre about 100,85,115 at 0.5XP.

15 Hide each viewport – Fig. 24.5.

16 Shade if required then save the model.

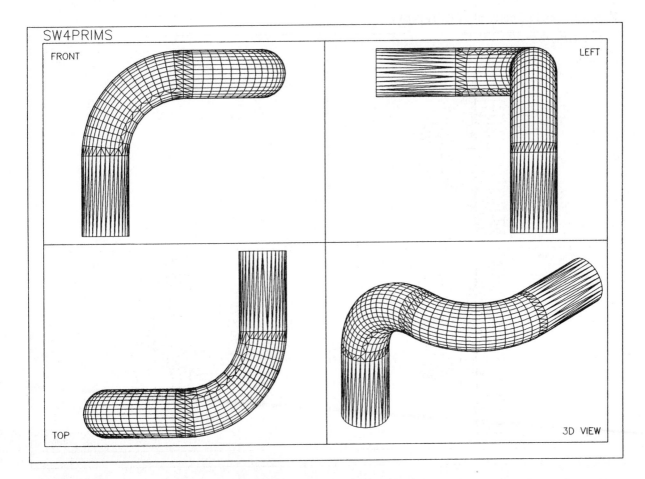

Figure 24.5 Circle extruded along two paths.

Revolved solids

Solids can be created by revolving objects (closed polylines, polygons, circles, ellipses, closed splines, regions) about the *X*- or *Y*-axes by a specified angle. As with the extrusions, very complex solids can be obtained from relatively simple shapes.

Revolved model 1: the moulding

1 Open SOLA3 with layer MODEL current, UCS BASE and with the lower left viewport active.

2 Using the moulding sizes from Fig. 24.1, create a closed polyline shape, the origin being at the point 0,0.

3 Zoom centre about the point –35,0,0 at 0.9XP in all viewports.

4 From the menu bar select **Draw–Solids–Revolve**
   prompt   Select objects
   respond pick the polyline then right-click
   prompt   Axis of revolution – Object/X/Y ...
   enter   **X** <R> – the X axis option
   prompt   Angle of revolution<full circle>
   respond right-click for full circle

5 The moulding shape is displayed as a revolved solid

6 Hide to give Fig. 24.6.

7 Save model if required as RV1PRIMS.

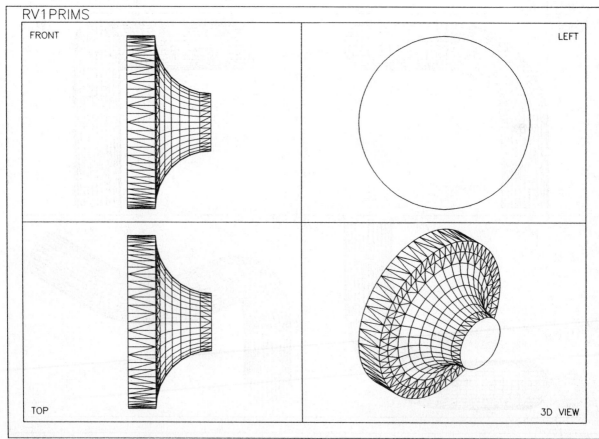

Figure 24.6 The moulding as a revolved solid.

Revolved model 2: a coupling of sorts

1 Open SOLA3 and refer to Fig. 24.7.

2 Ensure UCS BASE, layer MODEL current, lower left viewport active.

3 Draw the given shape as a closed polyline, the start point being at 20,–20. The actual shape design is at your discretion, the only requirement is that the length should not exceed 100 and the width should not exceed 80.

4 Zoom centre about 0,0,0 at 0.5XP in all viewports.

5 Select the Revolve icon from the Solids toolbar and:
   prompt Select objects
   respond pick the polyline then right-click
   prompt Axis of revolution ...
   enter   **X** <R>
   prompt Angle of revolution ...
   enter   **270** <R>

6 Hide to display the revolved solid as Fig. 24.7(A).

7 Undo the Hide and Revolved commands and REGENALL.

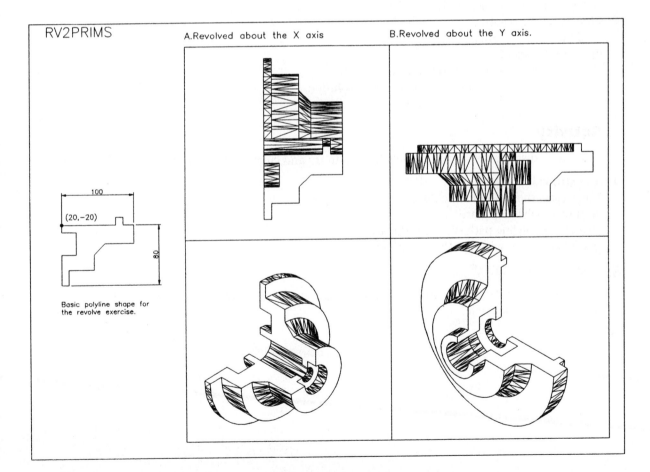

Figure 24.7 Revolved 'coupling' with specified angle of revolution.

172 *Modelling with AutoCAD*

8 At the command line enter **REVOLVE** <R>
   prompt   Select objects
   respond   pick the polyline then right-click
   prompt   Axis of revolution ...
   enter   **Y** <R>
   prompt   Angle of revolution ...
   enter   **270** <R>

9 Hide the model – Fig. 24.7B

10 Save if required as RV2PRIMS.

Summary

1 Swept solids are obtained with the extrude and revolve commands.
2 The two commands can be activated by icon, from the menu bar or by direct keyboard entry.
3 Very complex solids can be obtained with the two commands.
4 Only certain 'shapes' can be extruded/revolved. These are closed polyline shapes, circles, ellipses, polygons, closed splines or regions (more on this later).
5 Objects can be extruded:
   a) to a specified height
   b) along a path curve
   c) with/without a taper angle.
6 Objects can be revolved:
   a) about the *X*- or *Y*-axis
   b) about an object
   c) with a specified angle of revolution
   d) by specifying two points on the axis of revolution.

Activity

Tutorial 21 displays two swept models for you to try and create. The models are:

1 An extruded arch.
   Fairly easy, the procedure being:
   a) draw a polyline square
   b) create a polyline path curve for the arch
   c) care is needed with the UCS positions.

2 A revolved 'something'.

   Very easy as the design is your own and you can decide how you want the axis of revolution and the angle.

Chapter **25**

Boolean operations and composite solids

The basic and swept solids so far created are generally called **primitives** and are the 'basic tools for solid modelling'. With these primitives the user can create **composite solids**, so called because they are composed of two or more solid primitives, i.e.

a) primitive: a box, wedge, cylinder, extrusion, etc.
b) composite: a solid made from two or more primitives.

Composite solids are created from primitives using the three Boolean operations of union, subtraction and intersection. Figure 25.1 illustrates these operations using two primitives:

1 a box

2 a cylinder 'penetrating' the box.

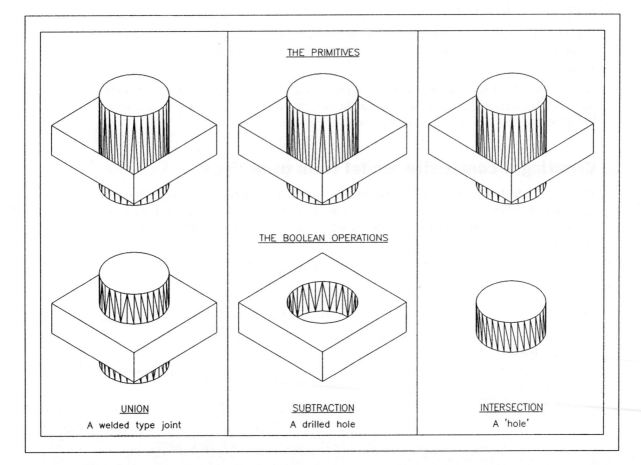

Figure 25.1 The three Boolean operations.

174 *Modelling with AutoCAD*

Union

This operation involves joining two or more primitives to form a single composite, the user selecting:

1 all objects to be unioned

The operation can be considered similar to welding two or more components together.

Subtraction

Involves removing one or more solids from another solid thereby creating the composite. The user selects:

1 the source solid

2 the solids to be subtracted from the source.

The result of a subtraction operation can be likened to a drilled hole, i.e. if the cylinder is subtracted from the box, a hole will result in the box.

Intersection

This operation gives a composite solid from other solids which have a common volume, the user selecting:

1 all objects which have to intersect.

The box/cylinder illustration of the intersection operation in Fig. 25.1 gives a 'disc shape' or 'hole', i.e. of the box and cylinder are intersected, the common volume is the disc shape.

Note: in Fig. 25.1 I have shown both the primitives and composites which result from the three Boolean operations. The plot has been with HIDE, and the user should note the appearance of the primitives compared to the composites.

Creating a composite model from primitives

There is no 'ideal way' to create a composite, i.e. the Boolean operations selected by one user may be different from those selected by another user, but the final result may be the same. To demonstrate this, we will create a simple L-shaped composite by three different methods, so:

1 Open SOLA3 with layer MODEL current and refer to Fig. 25.2.

2 In paper space erase all text and the four existing viewports.

Using Floating Viewports, create a single viewport from 10,10 to 370,260.

3 Return to model space and with UCS BASE create a box primitive with:
a) corner: 0,0,0
b) cube: length 100
c) colour: red.

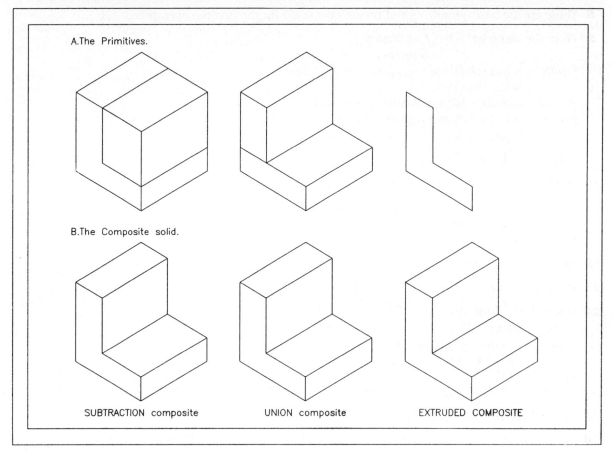

Figure 25.2 Creating a composite solid by three different methods.

4 Create another two boxes:
   a) corner: 125,125,0   *b*) corner: 125,125,30
       length: 100   length: 40
       width : 100   width : 100
       height: 30   height: 70
       colour: red   colour: blue

5 Restore UCS FRONT and create a polyline shape:
   From   500,250
   To   @100,0
   To   @0,30
   To   @−60,0
   To   @0,70
   To   @−40,0
   To   close

6 Restore UCS BASE and zoom centre about 100,250,0 at 0.6XP.

7 Create another Box:
   corner: 100,100,100
   length: −60
   width : −100
   height: −70
   colour: blue

176 *Modelling with AutoCAD*

8 These are the basic primitives and we are now ready for the Boolean operations.

9 From the menu bar select **Construct**
    Subtract
prompt   Select solids and regions to subtract from ...
           Select objects
respond pick large left red box then right click
prompt   Select solids and regions to subtract ...
           Select objects
respond pick large left blue box then right-click

10 The blue box is subtracted from the red box.

11 From the menu bar select **Construct–Union** and:
prompt   Select objects
respond pick the red and blue central boxes then right-click.

12 The two boxes are unioned.

13 Restore UCS FRONT.

14 Select the Extrude icon and:
prompt   Select objects
respond pick the red L shape then right-click
prompt   Path/<Height ...
enter   –100 <R>
prompt   Extrusion angle ... and enter 0 <R>

15 The L-shape polyline is extruded into a composite.

16 Now HIDE and the same L-shaped composite is obtained from three different operations as Fig. 25.2.

Note: my Fig. 25.2 has the original primitives as well as the composites – note the hide effect on these!

17 SHADE and note the difference in blue with the subtracted and unioned solids. Any comments in why this is?

18 From the Standard toolbar select the Mass Property from the List flyout and:
prompt   Select objects
respond pick the subtraction composite
prompt   text screen with details of:
           Mass
           Volume
           Bounding Box
           Centroid
           Moments of Inertia, etc.

19 Study the list then exit the command accepting the <N> default at the 'write to file' prompt.

20 Use the Mass Property icon and pick the other two composites and check if the table of information is the same. They should be with the exception of the bounding box.

21 *Questions*

 a) The mass and volume are both listed as 580000. Why is this?

  Release 13 assumes that the material density is 1. Hence mass will always be the same as volume.

  Release 12 users may be confused with this, but R13 does not support a materials library.

 b) Is the 580000 volume correct for our model?

Now that we have investigated the three Boolean operations, we are now ready to create composite solids which are more complex.

Chapter 26
Composite model 1 – a machine support

In this exercise we will create a composite solid from the box, wedge, cylinder and cone primitives and all three Boolean operations will be used. Once created, we will dimension the composite using viewport specific layers. The exercise is quite simple and you should have no difficulty in following the steps in the model construction. Try and reason out the coordinate information as it is given, i.e. do not simply enter the values without thinking about them.

1 Start AutoCAD R13 and begin a **new** drawing and:
prompt   New drawing dialogue box
respond   *a*) enter new drawing name as: **\R13MODEL\MACHSUPP=\R13MODEL\SOLA3**
              b) pick OK

2 Model space with layer MODEL current, lower right viewport active and UCS BASE. Refer to Fig. 26.1.

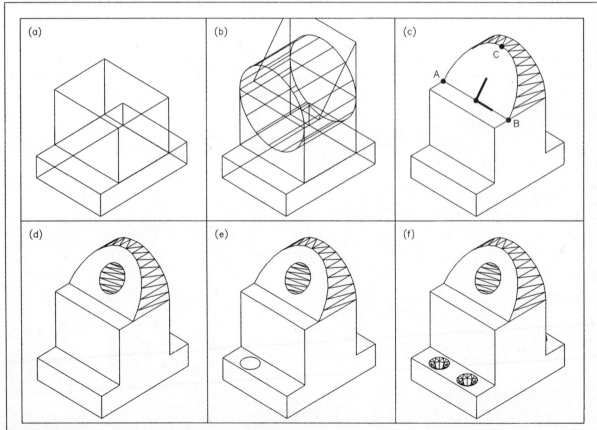

Figure 26.1 Creation of the solid composite MACHSUPP.

Composite model 1 **179**

3 Zoom centre about 35,50,50 at 150 magnification in all viewports.

4 Using the Box icon, create two box primitives:
    a) corner: 0,0,0   *b*) corner: 70,75,20
        length: 70   length: −70
        width : 95   width : −55
        height: 20   height: 50
        colour: red   colour: blue

5 This gives fig. (a).

6 Create a cylinder on top of the blue box with:
centre: 35,20,70
radius: 35
other end: enter **C** <R> the **@0,55,0** <R>
colour: green

7 Create a wedge with:
corner: 0,75,70
length: 40
width: 70
height: 50
colour: yellow

8 Rotate this yellow wedge about the point 0,75,70 by −90 to give fig. (b).

9 Select the Intersection icon from the Explode flyout of the Modify toolbar and:
  prompt   Select objects
  respond   pick the green cylinder and yellow wedge then right-click

10 Select the Union icon from the flyout and:
  prompt   Select objects
  respond   pick the red and blue boxes and the intersected part of the model then right-click

11 The composite at this stage is displayed in fig. (c).

12 Referring to fig. (c), set and save a new UCS position using the 3pt option with:
    a) origin point – MIDpoint of AB
    b) X-axis – ENDpoint B
    c) Y-axis – QUADrant C
    d) Save as SLOPE.

13 UCS should move? Remember UCSICON-A-OR

14 Create a cylinder with:
centre: 0,20,0
radius: 12
height: −50
colour: magenta

15 Now subtract the magenta cylinder from the composite – fig. (d).

16 Restore UCS BASE.

17 Create two further primitives:
    a) Cylinder   *b*) Cone
        centre: 20,10,0   centre: 20,10,20
        radius: 5   radius: 8
        height: 20   height: −5
        colour: cyan   colour: cyan

180 *Modelling with AutoCAD*

18 Union the two cyan primitives – fig. (e).

19 Rectangular array the cyan part:
   a) for 2 rows and 2 columns
   b) row distance: 75
   c) column distance: 30.

20 Using the Subtraction icon:
   a) pick the cuboid model then right-click
   b) pick the four cyan parts then right-click.

21 The model is now complete – fig. (f).

22 Hide then shade and you will realize the benefit of making each primitive a different colour?

23 REGENALL all and at this stage File–Save As. Is the name MACHSUPP?

Creating viewport specific layers

We now have a four viewport configuration of the solid composite and want to add some dimensions. These dimensions must be added on viewport specific layers which should be obvious to the user by now?

To create these layers:

1 Select from the menu bar **Data**
   Viewport Layer Controls
   New Freeze
   prompt   New viewport frozen layer name(s)
   enter   **DIMTL,DIMTR,DIMBL** <R>
   prompt   ?/Frozen/ ...
   respond right-click

2 With the top left viewport active, select **Data–Layers** from the menu bar and with the Layer Control dialogue box:
   a) note three new layer names with CN
   b) pick DIMTL line – turns blue
   c) pick **CurVP: Thw**
   d) C removed from DIMTL line
   e) pick OK.

3 With the top right viewport active use the layer control dialogue box and:
   a) pick DIMTR line
   b) pick CurVP: Thw
   c) pick OK.

4 Finally use the layer control dialogue box again and:
   a) pick DIMBL line
   b) pick CurVP: Thw
   c) pick OK.

5 What has been achieved in this section?
   a) we have made three new viewport specific layers for adding dimensions in three viewports
   b) these layers have been named DIMTL for the top left viewport, DIMTR for the top right and DIMBL for the bottom left

c) the layers were originally created frozen, and their states were:
   N: new viewport layer frozen
   C: currently frozen in the viewport
d) each layer has been currently thawed in a specific viewport, i.e. DIMTL is thawed in the top left viewport, but layers DIMTR and DIMBL are currently frozen in this viewport.

Adding the dimensions

1 With the lower left viewport active:
   a) restore UCS BASE
   b) make layer DIMBL current.

2 In paper space zoom in on the lower left viewport then return to model space.

3 Activate the Dimensioning toolbar and position it to suit.

4 Select the Linear Dimensioning icon and add the four dimensions in Fig. 26.2.

5 With the top left viewport active:
   a) restore UCS FRONT
   b) make layer DIMTL current
   c) add the two dimensions.

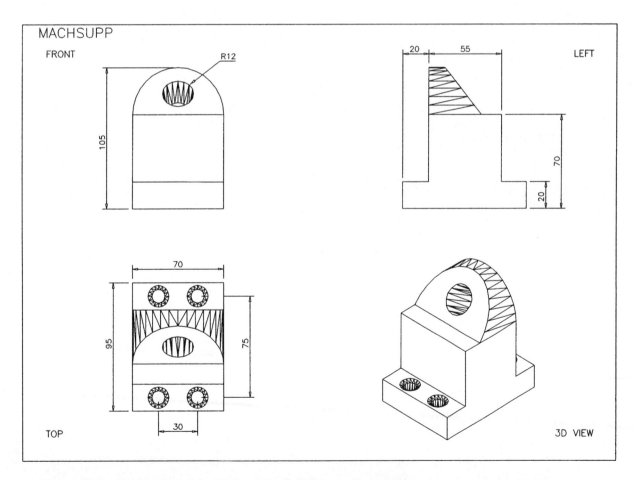

Figure 26.2 Complete model with dimensions of the machine support.

182 *Modelling with AutoCAD*

6 With the top right viewport active:
a) restore UCS LEFT
b) make layer DIMTR current
c) add the four dimensions.

7 The composite model is now complete – Fig. 26.2, and can be plotted with the VP layer frozen for effect.

8 This exercise should reinforce the benefit of viewport specific layers.

9 *Note*: I altered values in my DIMSOLID dimension style as follows:
a) arrowheads: 4
b) fit format: Text and Arrows
c) text height: 4.

10 The exercise is now complete. The composite was created from 'basic primitives' and used all three Boolean operations.

Chapter 27

Composite model 2 – a backing plate

In this exercise we will create a solid from an extruded swept primitive and then subtract various holes to complete the composite. The exercise will also entail altering the viewport layout of the SOLA3 standard sheet, which is an interesting exercise in itself. As with all the exercise do not just accept the entries – work out why the various values are being used.

The model

Refer to Fig. 27.1 which details the model to be created and gives all the relevant sizes. As an aside draw the three orthographic views as given and then add the isometric view. Time how long this takes. I spent about an hour to complete the four views with dimensions.

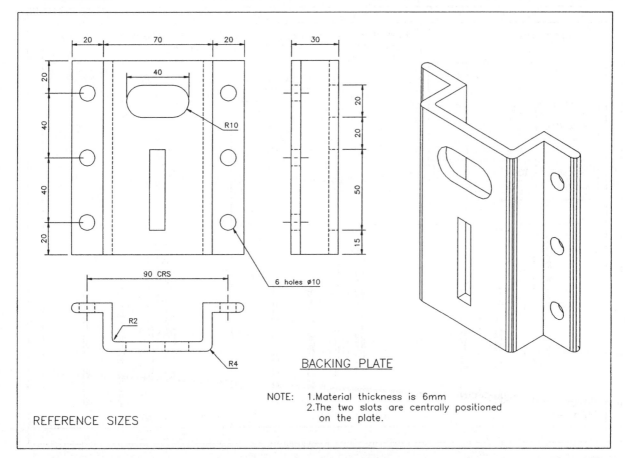

Figure 27.1 Backing plate drawn as orthographic and isometric.

184 Modelling with AutoCAD

Setting the viewports

1 Open your SOLA3 standard sheet and in paper space erase the four existing viewports and the text items. Refer to Fig. 27.2.

2 Make layer VP (yellow) current.

3 From the menu bar select **View**
                            Floating Viewports
                            1 Viewport
prompt   First point and enter **10,10** <R>
prompt   Other corner and enter **160,200** <R>

4 Create another three single floating viewports using the following co-ordinate inputs:
   a) first point: 160,10
      other corner: 310,70
   b) first point: 160,70
      other corner: 310,260
   c) first point: 310,70
      other corner: 370,260.

5 In model space ensure UCS BASE is current.

6 Set the 3D Viewpoint Presets as fig. (a).

7 In each viewport, zoom centre about 0,10,60 at 1XP.

8 Refer to fig. (a) and make viewport B active with layer MODEL current.

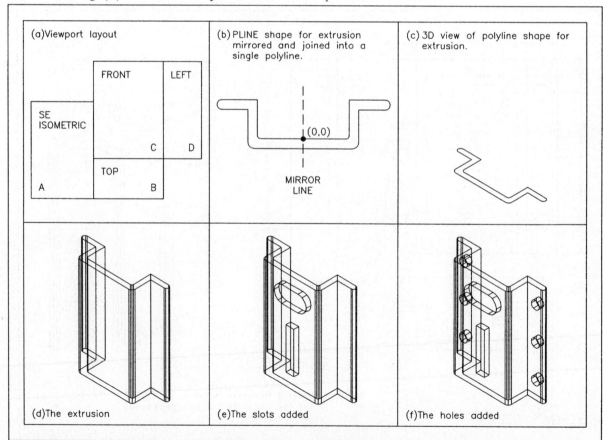

Figure 27.2   Steps in tyhe creation of the backing plate composite.

Creating the shape for the extrusion

1 Using the polyline icon create the following single polyline:

From point	0,0
To point	@27,0
To point	Arc endpoint @2,2
To point	Line endpoint @0,22
To point	@23,0
To point	Arc endpoint @0, −6
To point	Line endpoint @−17,0
To point	@0,−20
To point	Arc endpoint @−4,−4
To point	Line endpoint @−31,0
To point	right-click.

2 Mirror the polyline shape about a vertical line through 0,0.

3 Using **Modify–Edit Polyline** from the menu bar:
a) pick the right-hand polyline
b) enter **J** <R> – the join option
c) pick the left-hand polyline then right-click
d) enter **X** <R>.

4 The two 'halves' of the polylines have been joined into one.

5 The polyline shape is shown in fig. (b) in plan view and fig. (c) in 3D.

Completing the model

1 Make viewport A current.

2 At the command line enter **ISOLINES** <R> and:
prompt   New value for ISOLINES<24>
enter   **6** <R>.

Note: the ISOLINES value is being reduced to 6 due to the corners of the model. When extruded the value of 24 will result in these corners being 'very dense'.

3 Select the Extrude icon and:
prompt   Select objects
respond   pick the polyline then right-click
prompt   Path/<Height ... and enter **120** <R>
prompt   Extrusion taper angle and enter **0** <R>

4 The polyline is extruded – fig. (d).

5 Restore UCS FRONT.

6 Create a box primitive with:
corner: −5,15,0
length: 10
width : 50
height: 6
colour: blue

186 *Modelling with AutoCAD*

7 Create the following three primitives:

Box	*Cylinder*	*Cylinder*
corner: −10,85,0	centre: −10,95,0	centre: 10,95,0
length: 20	radius: 10	radius: 10
width : 20	height: 6	height: 6
height: 6	colour: green	colour: green
colour: green		

8 Union the three green primitives.

9 Subtract the blue and green primitives from the red extrusion by:
a) pick the red solid then right-click
b) pick the blue and green solids then right-click.

10 The model at this stage is displayed in fig. (e).

11 Make the first cylindrical hole with:
centre: 45,20,−24
radius: 5
height: 6
colour magenta

12 Rectangular array the magenta cylinder:
a) for 3 rows and 2 columns
b) row distance: 40
c) column distance: −90.

13 Now subtract the six magenta cylinders from the composite with viewport C active – probably easier.

14 The composite has been created – fig. (f).

15 The complete four viewport layout of the model is displayed in Fig. 27.3. Note that the hide effect does not 'show much' in two of the viewports.

16 Hide and shade your model in the 3D viewport. Again you should appreciate the colour effect of individual primitives.

17 Save your model as BCKPLATE.

18 How long did this exercise take, compared to the traditional orthographic/isometric drawing.

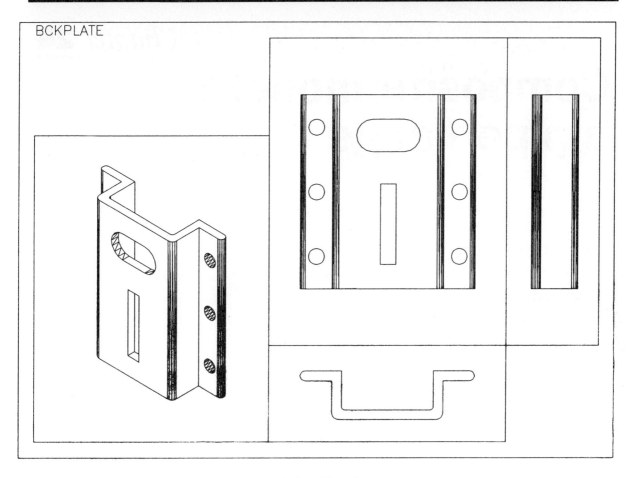

Figure 27.3 Complete solid composite model of backing plate.

Investigating the model

Before leaving this exercise enter at the command line **MASSPROP** <R> and:
prompt   Select objects
respond pick the composite then right-click
prompt   Text screen with a list of the composite mass properties
check   Mass: 102102.52 – my value!
respond <R><R>, i.e. no save to file.

This completes the backing plate exercise.

Chapter **28**

Composite model 3 – a pipe flange

This exercise will involve creating a solid composite mainly as a revolved swept primitive. Various primitives will be subtracted from the revolved object to complete the composite.

1 Open SOLA3 with layer MODEL current.

2 With UCS BASE, zoom centre about the point −100,−30,0 at 0.4XP in all viewports.

3 With the lower right viewport active, restore UCS RIGHT.

4 Draw two circles, centre at 0,0 with radii of 30 and 40.

5 Draw a line: from −200,0
                          to @0,100

6 Select the Revolve icon and:
 prompt   Select objects
 respond   pick the smaller circle then right-click
 prompt   Axis of revolution ...
 enter   **O** <R> – the object option
 prompt   Select an object
 respond   pick lower end of the line
 prompt   Angle of revolution
 enter   **70** <R>

7 Change the colour of the revolved pipe to green.

8 Revolve the larger circle, using the same entries as before. This pipe is to be blue.

8 Subtract the green pipe from the blue pipe, an operation which may require you to 'zoom in' on an area of the model.

9 Restore UCS BASE.

10 Draw a polyline:
 From   0,−30
 To   @0,−10
 To   @10,0
 To   Arc endpoint @10,−10
 To   Line endpoint @0,−50
 To   @30,0
 To   @0,70
 To   close

11 Using the Revolve icon:
 a) pick the polyline then right-click
 b) enter **X** <R> at the axis prompt
 c) enter **360** <R> at the angle prompt.

12 Change the colour of this revolved solid to magenta.

13 Erase the construction line.

14 With the top right viewport active, restore UCS LEFT.

15 Create a cylinder:
centre: 0,75,–50
radius: 10
height: 30
colour: yellow.

16 Polar array the yellow cylinder about the point 0,0 for six items.

17 Subtract the six cylinders from the magenta flange.

18 Still in the top right viewport with UCS LEFT, draw a polyline:
From   70,90
To   @30<–120
To   Arc endpoint @30<150
To   Line endpoint @30<60
To   close

19 Move this red polyline:
From   0,0
By   @0,0,–50

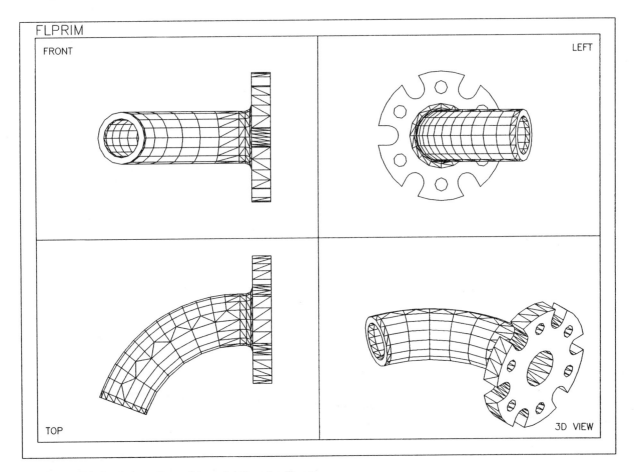

Figure 28.1 Composite solid model 3 – pipe/flange.

190 *Modelling with AutoCAD*

20 With the Extrude icon:
    a) pick the red polyline then right-click
    b) enter 30 as the height
    c) enter 0 as the taper angle.

21 Change the colour of the extrusion to cyan.

22 Polar array the cyan shape for six items about the point 0,0 with rotation.

23 Subtract the six cyan shapes from the flange.

24 Union the flange and pipe sections to create the single composite.

25 Hide each viewport to give a display as Fig. 28.1.

26 Try shading the model which is now complete.

27 Save your work as FLPRIM.

Chapter **29**

Modifying solids – fillet and chamfer

In this chapter we will investigate how a solid can be filleted and chamfered by considering the box and cylinder primitives and then a composite.

Box solid

1 Open SOLA3 standard sheet with the lower left viewport active, layer MODEL current and UCS BASE.

2 Use the Box icon to create a cube with:
corner: 0,0,0
length: 100

3 Zoom centre about 50,50,50 at 0.75XP in the 3D viewport and 1XP in the other three viewports.

4 Select the Chamfer icon and:
prompt	Polyline/ .../<Select first line>
respond	**pick a line on top surface**
prompt	Select base surface
and	*a*) one face of cube will be highlighted
	b) it will be a 'side' or the 'top'
	c) prompt is Next/<OK>
respond	*a*) right-click if top surface highlighted
	b) enter N<R> until top surface highlighted then right-click.
prompt	Enter base surface distance<?>
enter	**15** <R>
prompt	Enter other surface distance<?>
enter	**25** <R>
prompt	Loop/<Select edge>
respond	*a*) pick any three edges of top surface
	b) right-click

5 The top surface of the cube will be chamfered at the three selected edges.

Note: entering **L** <R> for loop, will allow all edges to be chamfered with the one pick.

6 Select the Fillet icon and:
prompt	Polyline/ ... /<Select first object>
respond	**pick a line on base surface of cube**
prompt	Enter radius
enter	**15** <R>
prompt	Chain/Radius/<Select edge>
respond	**C** <R> – the chain option
prompt	Edge/Radius/<Select edge chain>
respond	**pick the four edges of base** then right-click

192 *Modelling with AutoCAD*

7 The base is filleted – note corner effect.

8 In paper space, zoom previous then return to model space.

9 Hide each viewport to display the model as Fig. 29.1(a).

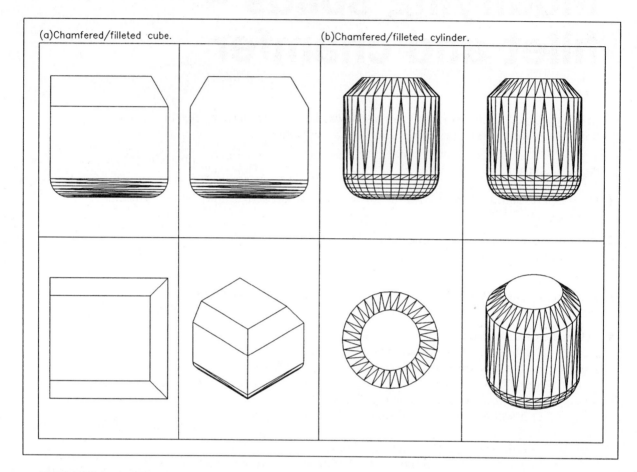

Figure 29.1 Modifying solids.

Cylinder solid

1 Erase the cube solid.

2 At the command line enter **ISOLINES** and check the value is 24.

3 Use the Cylinder-centre icon to create a solid with:
centre: 0,0,0
radius: 40
height: 100.

4 Zoom centre about 0,0,50 with the same XP values as before.

5 In paper space zoom the 3D viewport then return to model space.

6 Select the Chamfer icon and:
prompt   Polyline/ .../<Select first line>
respond   pick top surface circle edge
prompt   Select base surface
then   Next/<OK>

Modifying solids – fillet and chamfer **193**

respond	right-click as required surface is highlighted
prompt	Enter base surface distance
enter	**15** <R>
prompt	Enter other surface distance
enter	**15** <R>
prompt	Loop/<Select edge>
respond	**pick top circle edge** then right-click.

7 The top of the cylinder is chamfered.

8 Select the Fillet icon and:

prompt	Polyline/ ... /<Select first edge>
respond	**pick bottom circle**
prompt	Enter radius
enter	**20** <R>
prompt	Chain/Radius ...
respond	right-click as edge already highlighted.

9 Cylinder is filleted at the base.

10 In paper space zoom previous then return to model space.

11 Hide the model – Fig. 29.1(b).

12 Erase the model in preparation for the next exercise.

A composite solid

1 Create the following four primitives:

Box	*Box*	*Box*	*Cylinder*
corner: 0,0,0	corner: 30,0,30	corner: 100,0,30	centre: 90,50,0
length: 140	length: 40	length: 25	radius: 30
width: 100	width : 100	width: 100	height: 100
height: 100	height: 40	height: 40	colour: green
colour: red	colour: blue	colour: magenta	

2 Subtract the three coloured primitives from the red box.

3 Zoom centre about 70,50,50 at 1XP, 0.7XP in the 3D viewport.

4 In paper space zoom in on the 3D viewport then return to model space.

5 Using the Fillet icon:
  a) pick the top circle of green object
  b) enter radius of 20
  c) right-click

6 The top edge of the cylinder is filleted 'outwards' – in red!

7 Using the Chamfer icon:
  a) pick a long edge on top of the red box
  b) enter N<R> until top surface is highlighted
  c) right-click
  d) enter base surface distance of 20
  e) enter other surface distance of 20
  f) pick three long edges of top surface
  g) right-click.

8 Three edges of the top surface are chamfered.

9 *Task*:
   a) chamfer the fourth edge of the top surface of the red box, the two distances being 5
   b) chamfer the four near edges of the blue box, the two distances being 10
   c) fillet the four near edges of the magenta box, the radius being 10.

10 In paper space zoom previous then return to model space.

11 Hide the model in each viewport – Fig. 29.2.

12 The results of the chamfer/fillet operations are interesting?

13 Save the model if required.

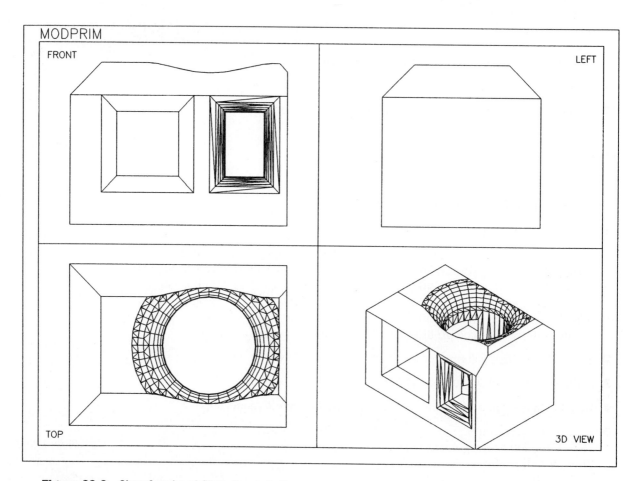

Figure 29.2 Chamfered and filleted composite.

Summary

1 Solids can be chamfered and filleted with the 'normal' commands.
2 Solids are chamfered/filleted:
  a) 'internally' if a primitive
  b) 'externally' if a 'hole'.
3 Individual edges can be chamfered/filleted.
4 The chamfer command has a LOOP option allowing a surface to be chamfered.
5 The fillet command has a CHAIN option allowing a surface to be filleted.
6 There are error messages which will be displayed if the chamfer distances or the fillet radius are too large for the model which is being modified. These messages are:
  a) blend of edge too great for local face curvature
  b) open end of spring curve not on boundary of face
  c) blend radius too big for adjacent face and edge curvature.
  These messages also give the prompts:
1 Failed to perform blend.
2 Failure while chamfer/fillet.

Chapter 30

Regions

A region is a closed 2D shape created from polylines, circles and arcs which can be used with the Extrude and Revolved commands to produce solid composites. When created, the region has certain characteristics:

- it is a solid of zero thickness
- it is confined to a plane
- it consists of **loops** – outer and inner
- the loops must be continuous closed shapes
- every region has one outer loop
- there may be several inner loops
- inner loops must be in the same plane as the outer loop
- they can be Extruded or Revolved
- they can be created using the Boundary command.

With the exception of the Boundary option, using regions does not allow the user any new ability to create solid models – it is another variation to be considered. We will investigate how regions are used with three examples.

Region example 1: the letter R

1 Open SOLA3 with the lower left viewport active and layer MODEL current. UCS set to BASE.

2 Refer to Fig. 30.1 and create the letter R from three **closed** polylines. The size of the letter is relatively unimportant, but I would suggest that the overall size is about 100 × 200.

This shape consists of one outer and two inner loops.

3 Select the Subtract icon and pick the outer loop and:

Message: No solids or regions selected.

4 Select the Region icon from the Rectangle flyout of the Draw toolbar and:
  prompt   Select objects
  respond **pick the three loops of the letter** then right-click
  prompt   3 loops extracted
           3 Regions created

5 With the Subtract icon:
  a) pick the outer loop then right-click
  b) pick the two inner loops then right-click.

6 The latter R has now been 'converted' into a region – fig. (a).

7 Zoom centre about the point 50,100,0 at 0.75XP

8 Select the Extrude icon and:
  a) pick the letter R then right-click
  b) enter a height of 50
  c) enter a taper angle of 0.

9 The letter is extruded as fig. (b).

10 Undo the extrusion with **U** <R>

11 Select the Extrude icon again and:
  a) pick the letter and right-click
  b) enter a height of 50
  c) enter a taper angle of 5
  d) hide the model – fig. (c).

12 Undo the hide and extrusion commands then select the Extrude icon for the last time and:
  a) pick the letter then right-click
  b) enter a height of 75
  c) enter a taper angle of –10
  d) hide to give fig. (d).

13 This completes the first example. Save?

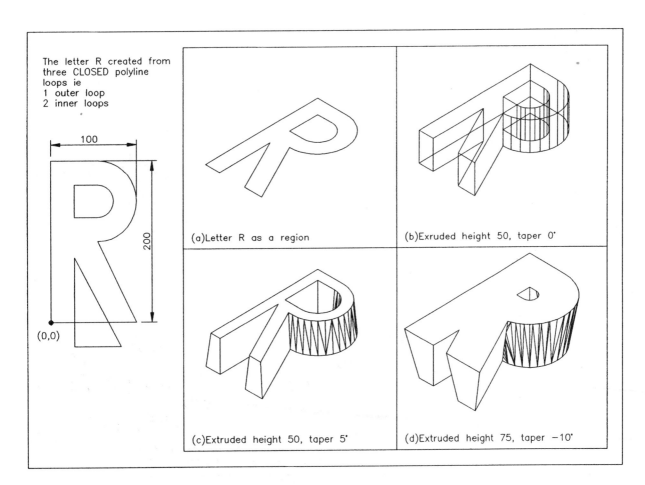

Figure 30.1 Region example 1 – the letter R.

198 Modelling with AutoCAD

Region example 2: a 'something'

1 Open SOLA3 layer MODEL, UCS BASE and lower left viewport active.

2 Refer to Fig. 30.2 and create the shape as shown from:
  a) a closed polyline
  b) four circles – any size and position
  c) use 25,25 as the start point.

3 Using the Region icon, pick all five objects then right-click and:
  prompt   5 loops extracted
        5 Regions created

4 *a*) Subtract the three loops (S) from the polyline
  b) Union the loop (U) to the polyline
  c) These operations give fig. (a).

5 Using the Revolve icon:
  a) pick the shape then right-click
  b) enter *X* as the axis of revolution
  c) enter –90 as the angle
  d) hide to give fig. (b).

6 Undo the hide and revolve commands to leave the original region.

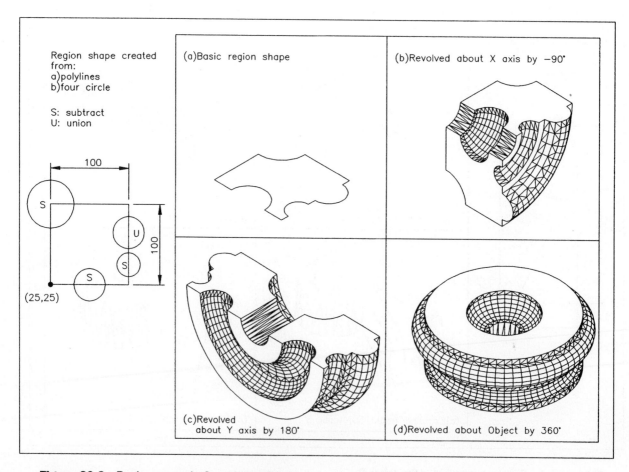

Figure 30.2 Region example 2 – a 'something'.

7 Using the Revolve icon again:
   a) pick the shape then right-click
   b) enter Y as the axis
   c) enter 180 as the angle
   d) hide to give fig. (c).

8 Undo the last operations to leave the region.

9 Draw a line from 0,0; to @0,0,100.

10 Select the 3D Rotate icon and:
   prompt   Select objects
   respond   pick the region then right-click
   prompt   Axis by ...
   enter   **X** <R>
   prompt   Point on X axis
   enter   **25,25,0** <R>
   prompt   Rotation angle
   enter   **90** <R>.

11 Select the Revolve icon and:
   a) pick the region then right-click
   b) enter **O** <R> for object
   c) pick vertical line
   d) enter 360 as the angle
   e) hide to give fig. (d).

12 This completes the exercise so save?

Region example 3: an intersected boundary shape

1 Open SOLA3, lower left viewport active, UCS BASE, layer MODEL current.

2 Refer to Fig. 30.3.

3 Create three intersecting circles with:
   a) centre 50,0; radius 50
   b) centre 0,50; radius 60
   c) centre 75,75; radius 75.

4 Create a new layer BND, colour blue and current.

5 Select the Boundary icon from the region flyout of Draw toolbar and:
   prompt   Boundary Creation dialogue box – Fig. 30.4
   respond   1 pick Object Type: Region
             2 pick Pick Points<
   prompt   Select internal point
   respond   pick point as indicated in Fig. 30.3
   prompt   Selecting everything ...
             Selecting everything visible ...
             Analyzing the selected data
   then   Select internal point
   respond   right-click
   prompt   1 loop extracted
             1 Region created
             BOUNDARY created 1 region.

200 *Modelling with AutoCAD*

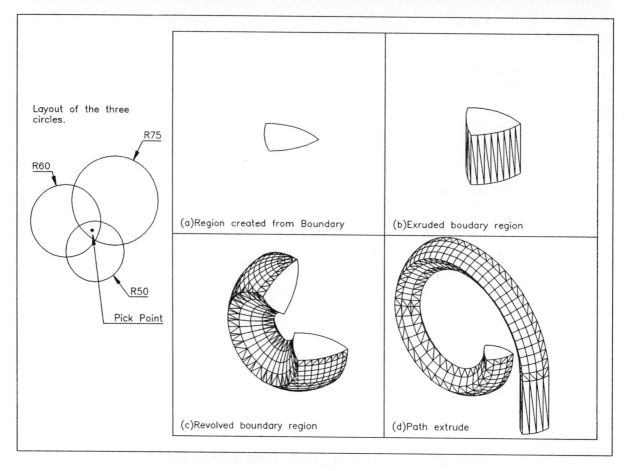

Figure 30.3 Region excample 3 – created from boundary

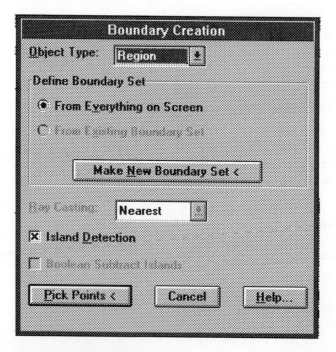

Figure 30.4 Boundary Creation dialogue box.

6 A blue shape is created and displayed. It is a region created from the Boundary command.

7 Erase the three red circles to leave the blue region – fig. (a).

8 Zoom centre about 0,0,0 at 0.5XP.

9 Using the Extrude icon:
a) pick the blue region then right-click
b) enter a height of 50
c) enter a taper angle of 0
d) hide – fig. (b).

10 Undo the last operations to leave the blue region.

11 Select the Revolve icon and:
a) pick the blue region then right-click
b) enter Y as the axis
c) enter 270 as the angle
d) hide to give fig. (c).

12 Undo the revolve operation to leave the blue region.

13 Restore UCS FRONT.

14 With layer MODEL current, draw a polyline:

From	0,0
To	@0,100
To	Arc endpoint @−200,0
To	Arc endpoint @120,0
To	right-click.

15 Zoom centre about −100,75,0 at 0.4XP and make layer BND current.

16 With the Extrude icon:
a) pick the blue region then right-click
b) enter P for the path option
c) pick the red polyline.

17 The blue region is extruded along the red path.

18 Hide to display the model as fig. (d).

19 Exercise is complete. Save?

Summary

1 A region is created from closed shapes made from polylines, circles arcs and ellipses.
2 Regions can be created using the Boundary command.
3 Regions can be extruded and revolved.
4 All parts of a region are extruded/revolved to the same height or angle.
5 Regions can be extruded along a path.

Chapter **31**

Moving solids

Solids can be moved and rotated with the traditional 2D and 3D commands. A command which is very useful with solids is **ALIGN**. We will demonstrate how this command can be used to re-align three solid primitives.

1. Open SOLA3 with layer MODEL current and the lower left viewport active. The UCS is BASE.
2. Create the following primitives:

Box	Cylinder	Wedge
corner: 0,0,0	centre: 120,120,0	corner: 100,–50,0
cube	radius: 50	length: 100
length: 100	height: 40	width: 60
colour: red	colour: green	height: 50
colour: blue		

3. Zoom centre about 100,100,0 at 0.5XP.

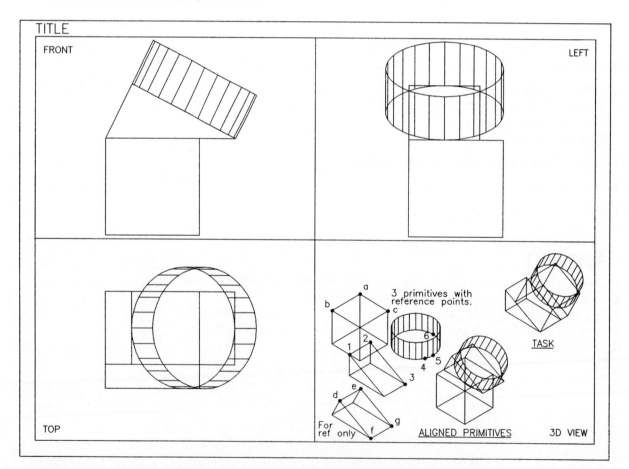

Figure 31.1 Aligning primitives.

4 The wedge is to be moved so that its sloped face is on the top surface of the cube, and then move the cylinder onto the wedge.

5 Refer to Fig. 31.1 (3D viewport) which gives reference points to assist with the various moves.

6 Set the running object snap to ENDPOINT.

7 Select the Align icon from the Modify toolbar and:
prompt	Select objects
respond	pick the blue wedge then right-click
prompt	1st source point and pick pt1
prompt	1st destination point and pick pta
prompt	2nd source point and pick pt2
prompt	2nd destination point and pick ptb
prompt	3rd source point and pick pt3
prompt	3rd destination point and pick ptc

8 The sloped face of the wedge will be aligned on the top face of the cube.

9 Cancel the endpoint running snap.

10 Note: this sequence includes a referenced wedge shape in Fig. 31.1 to assist you with the various pick points. You do not create this wedge.

Repeat the Align command and:
prompt	Select objects
respond	pick the green cylinder then right-click
prompt	1st destination point
respond	CENtre and pick base circle4
prompt	1st destination point
respond	MIDpoint and pick line de
prompt	2nd source point
respond	QUAdrant and pick base circle at pt5
prompt	2nd destination point
respond	ENDoint and pick pte
prompt	3rd source point
respond	QUAdrant and pick base circle at pt6
prompt	3rd destination point
respond	MIDpoint and pick line fg

11 The cylinder moves onto the wedge, but is positioned on the edge of the wedge. We want to reposition the cylinder at the 'centre' of the wedge face and we will use point filters for this.

12 Select the Move icon and:
prompt	Select objects
respond	pick the green cylinder then right-click
prompt	Base point ...
respond	CENtre and pick the base circle
prompt	Second point ...
respond	.Y point filter icon
then	MIDpoint and pick line eg
prompt	(need *XZ*)
respond	MIDpoint and pick line de

13 The green cylinder is positioned at the 'centre' of the wedgeface.

204 *Modelling with AutoCAD*

14 Hide each viewport then shade.

15 Save your model at this stage.

16 *Task*
  a) move the model to another part of the screen
  b) create another three primitives using the information previously given
  c) move the box onto the wedge then move the cylinder onto the box.

17 This completes this short chapter.

Chapter **32**

Slicing and sectioning solids

Solids can be sliced (cut) and sectioned relative to the three coordinate planes of the current UCS position. The two commands are very similar in their execution and when used:

a) the slice command results in a new composite model
b) the section command adds a region to the model

As usual we will investigate both commands by worked example.

Slice example 1

1 Open SOLA3 with layer MODEL current, lower right viewport active and UCS BASE. Refer to Fig. 32.1.

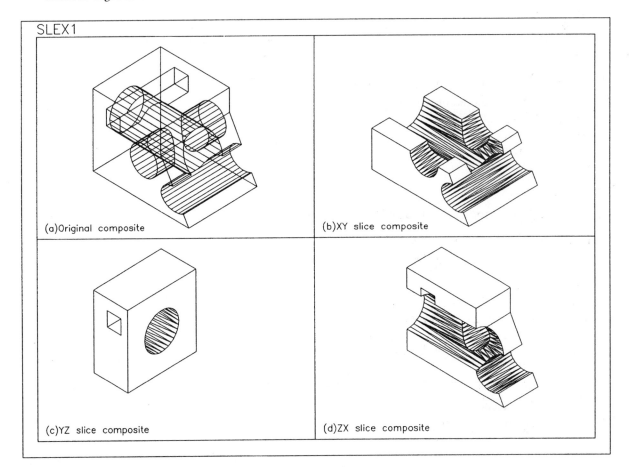

Figure 32.1 Slice example 1.

206 *Modelling with AutoCAD*

2 Create the following two primitives:

Box	*Wedge*
corner: 0,0,0	corner: 100,0,0
Cube	length: 40
length: 100	width: 100
colour: red	height: 70
	colour: yellow

3 Union the red and yellow primitives.

4 Zoom centre about the point 70, 50, 50 at 200 magnification in the 3D viewport and 150 magnification in the others.

5 Create a cylinder:

centre: 0,50,50
radius: 25
centre: other end @150,0,0
colour: blue

6 Subtract the blue cylinder from the composite.

7 Create three 'holes' with:

Box	*Cylinder*	*Elliptical cylinder*
corner: 20, 0, 65	centre: 75, 0, 50	centre: 120, 0, 25
length: 20	radius: 20	axis 1: @20, 0, 0
width: 100	other end at: @0,100,0	other axis: @0,15,0
height: 20	colour: green	other end at: @0,100,0
colour: green		colour: green

8 Subtract the three green 'holes' from the composite – fig. (a).

9 Set a new UCS origin point with **UCS** <R> and:

prompt	Origin/ ...
enter	**O** <R>
prompt	Origin point<0,0,0>
enter	**50,50,50** <R>

10 Save this UCS as MIDDLE.

11 At this stage save the model as **SLBOX** – we may use it again.

12 From the menu bar select **Draw**
Solids
Slice

prompt	Select objects
respond	pick the composite then right-click
prompt	Slicing plane by Object/Z axis/ ...
enter	**XY** <R> – the *XY* plane option
prompt	Point on *XY*-plane <0,0,0>
enter	**0,0,0** <R>
prompt	Both sides/<Point on desired side of the plane>
enter	**@0,0,–10** <R>

13 A new composite is displayed – fig. (b).

14 *Think about*: the three entries of XY; 0, 0, 0; @0, 0, –10.

15 Undo the slice command to restore the original composite – if problems open drawing SLBOX.

16 UCS still at middle?

17 Select the Slice icon from the Solids toolbar and:
 a) pick the composite then right-click
 b) enter YZ as the slicing plane
 c) enter 0,0,0 as a point on the plane
 d) enter @–10,0,0 as a desired point

18 A different composite is obtained – fig. (c).

19 Undo the last slice operation to restore the original composite then enter **SLICE** <R> at the command line and:
 a) pick the composite then right-click
 b) enter ZX as the slicing plane
 c) enter 0,0,0 as a point on the plane
 d) enter @0,10,0 as a desired point.

20 The resultant composite is displayed as fig. (d).

21 This completes the first example. Note that in Fig. 32.1, I have plotted the new composites on the one four viewport configuration.

 Any idea how this was obtained?

Slice example 2

1 Open SOLA3, layer MODEL, UCS BASE, lower right viewport active and refer to Fig. 32.2.

2 Create the following composites:

Box	*Sphere*
corner: 0, 0, 0	centre: 60, 60, 60
Cube	radius: 50
length: 120	colour: blue
	colour: 12

3 Zoom centre about the point 60, 60, 60 as 0.6XP in all viewports.

4 Subtract the blue sphere from the cube.

5 Create a red box with its **centre** at 60,60,60, cube length 30.

6 Do not union or subtract this box – leave it as a primitive.

7 Create a yellow cylinder with centre 45,45,120; radius 5 and height –40.

8 Rectangular array the yellow cylinder:
   a) for three rows and three columns
   b) row and column distances: 15.

9 Subtract the nine yellow cylinders from the composite – zoom needed? Remember to leave the red box as it is.

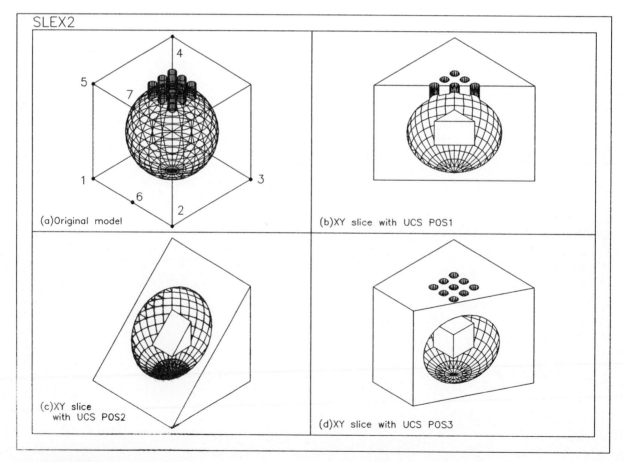

Figure 32.2 Slice example 2.

Slicing and sectioning solids 209

10 At this stage hide the model – nothing special – fig. (a).

11 Save the model as **NUCLEAR**.

12 Using the UCS three-point option, set and save three new UCS positions with the following information and referring to fig. (a):

		1	2	3
a)	origin:	pt1	pt1	pt6 – MIDpoint
b)	X axis:	pt3	pt2	pt3
c)	Y axis:	pt5	pt4	pt7 – MIDpoint
d)	name	POS1	POS2	POS3

13 Restore UCS POS1

14 Activate the slice command and:
   a) window the model then right-click
   b) enter XY as the slicing plane
   c) enter 0,0,0 as a point on the plane
   d) enter @0,0,–10 as a desired point
   e) model displayed as fig. (b).

15 Undo the last slice operation and restore UCS POS2.

16 Repeat step 14 using the same entries – fig. (c).

17 Undo the slice command and restore UCS POS3.

18 Again repeat step 14 – fig. (d).

Note

1 The first exercise created the sliced composites with the UCS set to one position (MIDDLE) and the three different slice planes options activated.

2 The second exercise created the sliced composite using only the *XY*-plane option but three different UCS positions.

3 Both methods are equally valid.

4 The SLICE (and SECTION) command has other options which I have not considered. These are Object, *Z*-axis and View. They are fairly easy to use and you can investigate them for yourself.

Section example

1 Open SOLA3, layer MODEL current and UCS BASE.

2 Refer to Fig. 32.3.

3 With the lower right viewport active, create two primitives:

Cylinder
centre: 0,0,0
radius: 50
height: 100
colour: red

Box
centre: 0,0,50
Cube
length: 60
colour: blue

4 Zoom centre about the point 0,0,50 at 1XP in three viewports but 0.75 in the 3D viewport.

5 Chamfer the top surface of the cylinder 10 × 10 and fillet the bottom surface of the cylinder 15 – easy!

6 Subtract the blue cube from the cylinder.

7 Create another two primitives:

Cone
centre: 0,0,100
radius: 25
height: −80
colour: green

Cylinder
centre: 0,0,50
radius: 10
other end at: @50,0,0
colour: yellow

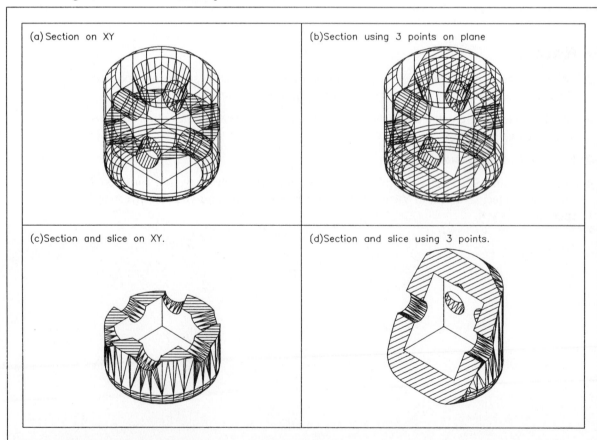

Figure 32.3 Section/slice example.

Slicing and sectioning solids

8 Polar array the yellow cylinder about the point 0,0 for six items with rotation.

9 Subtract the yellow and green primitives from the composite – zoom in paper space needed?

10 Hide and shade then regen. Save at this stage.

11 Make layer HATCH (cyan) current – or make a new hatch layer if required.

12 In paper space, zoom in on the 3D viewport then return to model space.

13 From the menu bar select **Draw**
                              Solids
                              Section
   prompt   Select objects
   respond   pick the composite then right-click
   prompt   Section plane by Object ...
   enter   XY <R>
   prompt   Point on XY plane <0,0,0>
   enter   **0,0,50** <R>

14 A cyan region is added to the model and may be quite difficult to 'see'. I changed my HATCH layer colour to 9.

Note: **no hatching is added to this region** – R12 users beware.

15 Set the UCS origin to 0,0,50.

16 At the command line, enter **HATCH** <R> and:
   prompt   Pattern ... and enter **U** <R>
   prompt   Angle ... and enter **45**
   prompt   Spacing ... and enter **4** <R>
   prompt   Double ... and enter **N** <R>
   prompt   Select objects
   enter   **L** <R><R> – two returns to select the region.

17 Hatching is added to the region as fig. (a).

18 Hide the model – no hatching?

19 At the command line enter:
   a) U<R> – undo the hatch command
   b) U<R> – undo the UCS move
   c) U<R> – undo the section command.

20 With UCS BASE, select the Section icon from the Solids toolbar and:
   prompt   Select objects
   respond   pick the composite then right-click
   prompt   Section plane by ...
   enter   **30, –30, 0** <R> – the three point option
   prompt   2nd point on plane
   enter   **–30, –30, 0** <R>
   prompt   3rd point on plane
   enter   **0, 30, 100** <R>

21 A region is added to the model but no hatching.

212 *Modelling with AutoCAD*

22 Set a UCS using the three point option with:
   a) origin at 0,−30,0
   b) X axis at 30, −30, 0
   c) Y axis at 0, 30, 100.

23 Using the BHATCH command, add hatching to this region using the same entries as step 16 – fig. (b).

24 *Note*: my fig. (a) and fig. (b) have been plotted without hide to enable the hatch effect to be 'seen'.

25 This completes the section exercise but do not exit.

Task

The slice and section commands can be combined to produce a hatched section component. Can you produce to effect shown in figs (c) and (d)? These displays use the same sequence of commands as before. I section first, then sliced. You need to be careful with the UCS.

Summary

1 The SLICE command produces a new composite.
2 The SECTION command adds a region to the model, but does not add any hatching to this region. Hatching must be added as a separate operation.
3 Both commands are very similar in execution and have several options:
   a) the *XY*, *YZ* and *ZX* slicing/sectioning planes
   b) relative to an Object
   c) using three points on the plane – by coordinate entry or by referencing existing entities
   d) other options.
4 The slice command requires:
   a) a point on the slicing plane
   b) a point on the desired side of the plane which 'is to stay'.
5 The section command only requires a point on the sectioning plane.
6 The position of the UCS is important with both commands.
7 The commands can be activated:
   a) by icons from the Solids toolbar
   b) from the menu bar with **Draw–Solids**
   c) by keyboard entry.

Chapter 33

A detailed drawing

In this chapter we will create a 'complex' composite and use it as a detailed drawing, i.e. we create sectional views, true shapes and add dimensions. The exercise will use all 'our solid modelling knowledge' and will make use of viewport specific layers.

The exercise is divided into sections, which should make the various sequences easier to follow.

Traditional 2D drawing

The component which will be used for the detailed drawing is a desk tidy and the relevant sizes are given in Fig. 33.1. As an exercise on its own, draw the given views as a traditional orthogonal drawing and time how long it took to complete. I think you will be surprised at the time required to complete this drawing, especially with the isometric being added. The true shape of the slope with the holes in it is also trickier than at first thought.

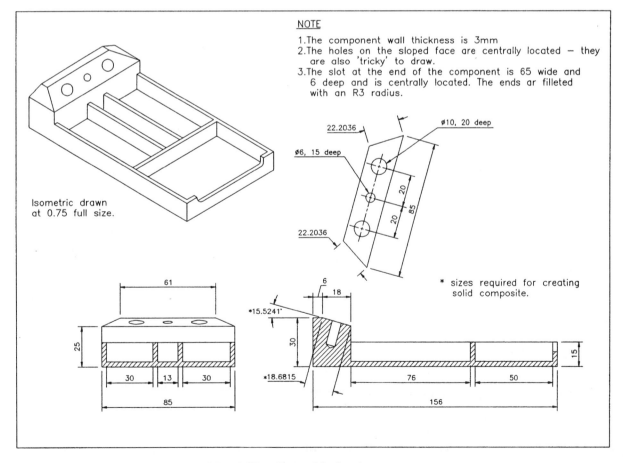

Figure 33.1 Desk tidy as a traditional 2D orthographic drawing.

214 *Modelling with AutoCAD*

There are two sizes from this drawing which are necessary for completing the solid model composite, this being:

1 The slope surface angle to the horizontal: 15.52411.

2 The slope surface width: 18.68154169.

The basic composite

1 Begin a new drawing with the New Drawing Name: \R13MODEL\TIDY=\R13MODEL\ SOLA3.

2 Layer MODEL current, lower right viewport active and UCS BASE.

3 In paper space select the STRETCH icon and:

prompt	Select objects
enter	**C** <R>
prompt	First corner and enter **150,100** <R>
prompt	Other corner and enter **250,150** <R>
prompt	4 found
	Select objects
respond	right-click
prompt	Base point and enter **190,135** <R>
prompt	Second point and enter **@25,25** <R>

4 The viewport layout has now been altered.

5 In model space, zoom centre about the point 75,38,12 at 0.75XP in the 3D viewport and 1XP in the other three viewports.

6 With the upper left viewport active, restore UCS FRONT.

7 Refer to Fig. 33.2.

8 Using the Polyline icon create the side of the model by drawing:

From	0, 0
To	@156, 0
To	@0, 15
To	@−132, 0
To	@0, 10
To	@−24, 0
To	close

9 Using the Extrude icon, extrude the red polyline for a height of **−85** with 0 taper – fig. (a).

10 Restore UCS BASE.

11 Create a box with its corner at 0,12,25 with length 6, width 61 height 5 and colour red.

12 Create three red wedges with:
   a) corner: 6,12,25; length 18; width 61; height 5
   b) corner: 6,12,25; length 12; width −6; height 5
   Rotate this wedge about the point 6, 12, 25 by **−90**
   c) corner: 6,73,25; length 12; width 6; height 5
   Rotate this wedge about the point 6,73,25 by **90**.

13 Union the three wedges and the box with the extrusion – fig. (b).

The detailed drawing **215**

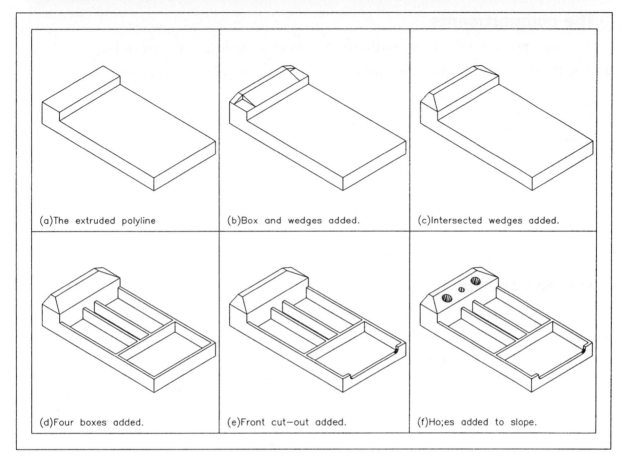

Figure 33.2 Steps in the construction of the desk tidy.

14 Create another two red wedges with:
  a) corner: 6,0,25; length 18; width 12; height 5
  b) corner: 24,12,25; length 12; width −18; height 5
    Rotate this wedge about the point 24,12,25 by **−90**.

15 From the menu bar select **Construct**
                  Intersection
  prompt Select objects
  respond pick the two newly created wedges then right-click

16 From the menu bar select **Construct**
                  3D Mirror
  prompt Select objects
  enter **L** <R><R> – yes two returns – why?
  prompt Plane by Object ...
  enter **ZX** <R>
  prompt Point on *ZX*-plane
  enter **0, 42.5, 25** <R>
  prompt Delete old objects and enter **N** <R>.

17 Union the two sets of intersected wedges with the composite.

18 The basic 'shape' of the desk tidy is now complete – fig. (c).

19 Save at this stage?

216 *Modelling with AutoCAD*

The compartments

These will be created from four boxes which will then be subtracted from the composite.

1 With the lower right viewport active and UCS BASE, create the following four boxes:

	box1	*box2*	*box3*	*box4*
corner:	153,3,3	100,3,3	100,36,3	100,52,3
length:	–50	–76	–76	–76
width :	79	30	13	30
height:	20	20	20	20
colour:	yellow	blue	green	blue

2 Subtract these four boxes from the composite – fig. (d).

3 Try hide and shade then regen.

4 Save again?

The end cut out

This will be obtained from an extruded polyline.

1 Lower-right viewport active with UCS BASE.

2 Set a new UCS position using the three-point option and:
   a) origin at 156, 0, 0
   b) X-axis at 156, 85, 0
   c) Y-axis at 156, 0, 15
   d) Save as NEWFRONT.

3 If the icon does not move to this new origin point then remember the UCSICON–ALL–OR sequence.

4 Draw a polyline:

From	10,15
To	@0,–3
To	Arc endpoint: @3,–3
To	Line endpoint: @59,0
To	Arc endpoint: @3,3
To	Line endpoint: @0,3
To	close

5 Extrude this polyline for a height of –3 with 0 taper.

6 Subtract the extruded polyline from the composite – fig. (e).

The holes on the slope

The holes will be created from cylinders, so:

1 Lower-right viewport active and restore UCS BASE.

2 In paper space zoom in on the slope surface then return to model space.

3 Set and save a new UCS on the slope with the three-point option and:
   a) origin at 24, 42.5, 25
   b) X-axis at 24, 85, 25
   c) Y-axis at 6, 42.4, 30
   d) Save as SLOPE.

The detailed drawing **217**

4 Create two primitives:

	Cylinder	*Cone*
centre:	0,9.34077,0	0,9.34077,–12
radius:	3	3
height:	–12	–3
colour:	magenta	magenta

5 Question: why 9.34077?

6 Union the magenta cylinder and cone then subtract from the composite.

7 Create another two cylinders:

	cyl1	*cyl2*
centre:	20,9.34077,0	–20,9.34077,0
radius:	5	5
height:	–20	–20
colour:	magenta	magenta

8 Subtract these two cylinders from the composite.

9 In paper space, zoom all then back to model space.

10 Restore UCS BASE.

11 Your model is now complete – fig. (f) and can be saved.

12 To check your accuracy, select the Mass Property icon from the Object Properties toolbar flyout and pick the composite and:
Volume: 106311.43
Centroid: 50.70, 42.50, 8.92
These were my figures!

Altering the viewport configuration

To complete the exercise, the viewport layout has to be altered so:

1 Enter paper space.

2 Select the MOVE icon and:
 a) pick the border of top-left (FRONT) viewport then right-click
 b) enter 215,160 as the base point
 c) enter 370,10 as the second point.

3 Repeat the MOVE command and:
 a) pick the top right (LEFT) viewport border
 b) enter 370,160 as the base point
 c) enter 105,10 as the second point.

4 MOVE the 3D viewport from:
 a) a base point of 370,10
 b) to the point 165,110.

5 MOVE the old lower left (TOP) viewport:
 a) from 10,10
 b) to 165,110.

6 In model space make the new lower right viewport active and:
 a) set the 3D Viewpoint Presets to Right
 b) zoom centre about 75,38,12 at 1XP.

218 *Modelling with AutoCAD*

7 With the top right viewport active:
   a) restore UCS SLOPE
   b) from the menu bar select **View**
                              3D Viewpoint Presets
                              Plan View
                              Current

8 The model is displayed in a 'vertical position'.

9 Select from the menu bar **View–3D Dynamic View**
   prompt   Select objects
   respond pick the composite then right-click
   prompt   CAmera/ ...
   enter   **TW** <R> – the twist option
   prompt   New view twist
   enter   **74.47589** <R> – why this figure?
   prompt   CAmera/ ...
   enter   **X** <R>

10 Restore UCS BASE.

11 Zoom centre about 75,38,12 at 1XP.

12 At this stage your viewport configuration is Fig. 33.3

13 Save – just in case?

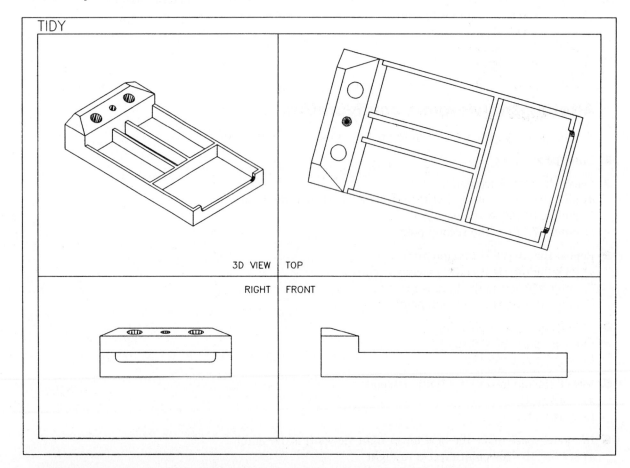

Figure 33.3 Desk tidy layout after viewport manipulation and the DVIE command.

The detailed drawing **219**

Viewport specific layers

To enable a true shape, sections and dimensions to be 'added' to the layout it is necessary to create several viewport specific layers, so:

1 From the menu bar select **Data**
Viewport Layer Controls
New Freeze

prompt	New Viewport frozen layer name(s)
enter	**TSHAPE** <R> – for the true shape 'extraction'
prompt	?/Freeze/Thaw ...
enter	**N** <R>
prompt	New Viewport frozen layer name(s)
enter	**SECTFRONT,SECTRIGHT** <R> – for the sections
prompt	?/Freeze/Thaw ...
enter	**N** <R>
prompt	New Viewport frozen layer name(s)
enter	**DIMTOP,DIMFRONT,DIMRIGHT,DIM3D** <R> – obvious?
prompt	?/Freeze/Thaw ...
respond	right-click

2 Using the layer control dialogue box:
a) note the new layers with:
 C: currently frozen in viewport
 N: newly frozen
b) change the three new DIM layers colour to magenta
c) change the two new SECT layers colour to suit
d) leave the TSHAPE layer colour.

3 Make the upper right viewport active, and from the menu bar select **Data–Layers**. Using the Layer Control dialogue box:
a) pick TSHAPE layer line
b) pick DIMTOP layer line
c) pick **Cur VP: Thw** – no C
d) pick OK.

4 With the lower right viewport active, use the layer control dialogue box and:
a) pick DIMFRONT layer line
b) pick SECTFRONT layer line
c) pick Cur VP: Thw
d) pick OK.

5 With the lower left viewport active use the layer control dialogue box again and:
a) pick DIMRIGHT and SECTRIGHT layers
b) pick Cur VP: Thw
c) pick OK.

6 Note: in this section we have created viewport specific layers and thawed certain named layers in specific viewports. The reason for this will become obvious as we complete the detail drawing – I hope!

The true shape

1 Make the upper right viewport active.

2 Make layer TSHAPE current and restore UCS BASE.

3 In paper space, zoom in on the top-right viewport then return to model space.

4 Using the LINE command draw on top of the four sides of the sloped face using ENDpoint, taking care to pick the correct endpoints. Add the three circles using CENtre and QUAdrant for the centre and radii selections.

5 With the layer control dialogue box:
a) pick MODEL layer
b) pick **Cur VP: Frz**
c) pick OK.

6 The composite will be frozen in the upper right viewport to leave the true shape of the sloped surface.

7 Paper space, zoom previous the model space.

The sections

1 Lower right viewport active with layer SECTFRONT current.

2 Restore UCS FRONT.

3 Select the Section icon and:
a) pick the composite then right-click
b) enter XY as the section plane
c) enter 0,0,–42.5 as a point on the *XY*-plane.

4 At the command line enter **HATCH** <R> and:
a) pattern and enter U <R>
b) angle and enter 45 <R>
c) spacing and enter 3 <R>
d) double and enter N <R>
e) select objects and enter L <R><R> – yes two – why?

5 Using **Data–Layers**:
a) pick MODEL layer
b) pick Cur VP: Frz
c) pick OK.

6 Only the section detail will be displayed in the lower right viewport. Some lines missing?

7 *a*) make the lower left viewport active
b) make layer SECTRIGHT current
c) restore UCS RIGHT.

8 Activate the SECTION command and:
a) pick the composite then right-click
b) enter *XY* as the section plane
c) enter 0, 0, 50 as a point on the plane.

9 Hatch this section using the same entry as step 4.

10 Using the layer control dialogue box:
a) pick layer MODEL
b) pick Cur VP: Frz then OK.

11 Section only displayed in the viewport, but again several lines 'behind' the section are missing.

12 Note: this section demonstrates the limitations of R13 with true shapes and sections.

Realigning the true shape viewport

The viewport containing the true shape is not aligned correctly with the front section viewport, and we will now rectify this. Refer to Fig. 33.4.

1 Enter paper space.

2 Select the MOVE icon and:
  a) select objects and pick the top right viewport border then right-click
  b) pick INTersection then ptA as the base point
  c) pick INTersection then ptB as the second point.

3 The complete viewport with the true shape will be moved.

4 Select the MOVE icon again and:
  a) enter **P** <R><R>
  b) enter 0,0 as the base point
  c) enter @50<74.4759 as the second point.

5 The true shape viewport is now aligned relative to the sloped surface from the front view.

6 Return to model space.

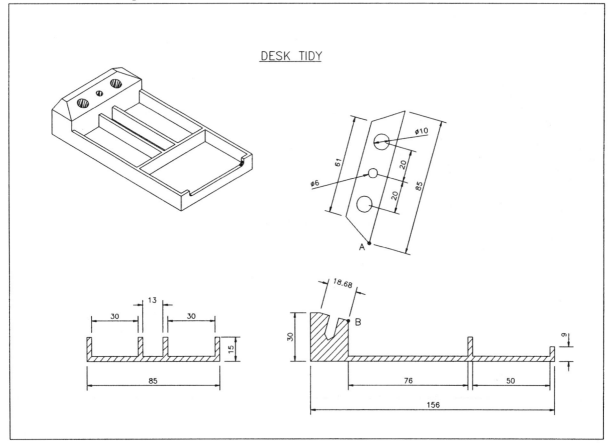

Figure 33.4 Completed detail drawing of the desk tidy.

Dimensioning the model

1 *a*) make lower right viewport active
   b) make layer DIMFRONT current
   c) restore UCS FRONT.

2 Check your dimension style and alter:
   a) text height: 3
   b) arrowheads: 3.

3 Add dimensions as required – refer to Fig. 33.4.

   Note: I altered the aperture box to 2 to allow ease of selection.

4 *a*) Make the upper right viewport active
   b) make layer DIMTOP current
   c) restore UCS SLOPE.

5 Dimension the true shape.

6 Add dimensions to the lower-left viewport, remembering layer and UCS.

And finally

The detail drawing is almost complete but:

1 Enter paper space.

2 Freeze layer VP.

3 Add title and your display should resemble Fig. 33.4. It can now be saved, and you can have a well-earned rest.

Summary

1 A detail drawing requires the use of viewport specific layers.
2 True shapes need to be drawn and cannot automatically be extracted
3 Section detail can be extracted but hatching is not added to this section.
4 R12 solid model users should note:
   a) there are no commands in R13 to extract features or profiles
   b) sections require hatching to be added by the user.
   c) viewports cannot display composite hidden detail due to the absence of the profile command
   d) why is this? The answer is that you need to use DESIGNER.

Chapter **34**

Solid model block assembly

This exercise will create an assembly from solid model blocks. We will also investigate interference and external references. The components of the assembly consist of two trays and four support legs and are fully detailed in Fig. 34.1.

The tray

1. Begin a new drawing with the name: **\R13MODEL\TRAY=\R13MODEL\SOLA3**.
2. Zoom centre about 55,40,20 at 0.75XP in all viewports.
3. Make a new layer TRAY, colour green and current, with UCS base.

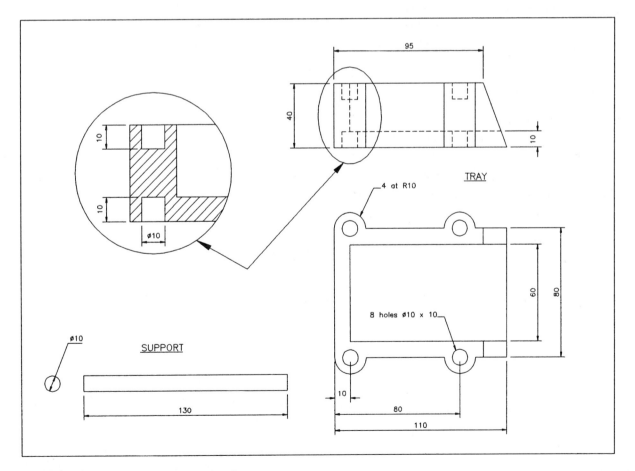

Figure 34.1 Tray and support details.

224 *Modelling with AutoCAD*

4 With the lower right viewport active, create two primitives from:

Box	Cylinder
corner: 0,0,0	centre: 10,0,0
length: 110	radius: 10
width: 80	height: 40
height: 40	

5 Rectangular array the cylinder:
  a) for two rows and two columns
  b) row distance: 80
  c) column distance: 70.

6 Union the four cylinders and the box – Fig. 34.1(a).

7 Create another two primitives:

Box	Wedge
corner: 10,10,10	corner: 110,0,40
length: 110	length: −15
width: 60	width: 80
height: 40	height: −40.

8 Subtract the box and wedge from the composite – fig. (b).

9 Create a cylinder, centre at 10,0,0, radius 5 and height 10.

10 From the menu bar select **Construct**
                                     3D Array
                                     Rectangular

prompt	Select objects
enter	**L** <R><R> – two returns to pick the cylinder
prompt	Number of rows and enter 2 <R>
prompt	Number of columns and enter 2 <R>
prompt	Number of levels and enter 2 <R>
prompt	Row distance and enter 80 <R>
prompt	Column distance and enter 70 <R>
prompt	Level distance and enter 30 <R>

11 Subtract the eight cylinders from the composite – paper space zoom needed? – fig. (c).

12 Lower-right viewport active and UCS BASE.

13 At the command line enter **BASE** <R> and:
prompt	Base point ...
enter	**0,0,0** <R>

14 From the menu bar select **File–Save** to save the model as TRAY in the R13MODEL directory.

The support

1 Begin a new drawing with the name: **\R13MODEL\SUPPORT=\R13MODEL\ SUPPORT**.

2 In model space:
a) zoom centre about 0,0,60 at 0.75XP in all viewports
b) lower right viewport active
c) UCS BASE.

3 Make a new layer – SUPPORT, colour blue and current.

4 Create a cylinder with:
centre: 0,0,0
radius: 5
height: 130 – fig. (d).

5 Enter **BASE** <R> at the command line and:
prompt   Base point ...
enter   **0,0,0** <R>

6 From the menu bar select **File–Save** to save the model as SUPPORT.

Insert exercise 1

1 Begin a new drawing with the name: **\R13MODEL\ASSEMBLY=\R13MODEL\ SOLA3**.

2 Zoom centre about 55,40,90 at 0.5XP in all viewports.

3 Make two new layers:
a) TRAY colour green and current
b) SUPPORT colour blue.

4 Display the External References toolbar.

5 Make the lower right viewport active with UCS BASE.

6 From the menu bar select **Draw**
                                Insert
                                Block ...
prompt   Insert dialogue box
respond   *a*) pick File ...
               b) pick the r13model directory name
               c) pick the tray.dwg file name
               d) pick OK
prompt   Insert dialogue box
respond   pick OK
prompt   Insertion point
enter   **0, 0, 0** <R>
prompt   X scale ... and enter 1 <R>
prompt   Y scale ... and enter 1 <R>
prompt   Rotation ... and enter 0 <R>

7 At this stage select **File–Save** to save the model as ASSEMBLY.

226 *Modelling with AutoCAD*

8 At the command line enter **INSERT** <R> and:
   prompt   Block name (or ?)<TRAY>
   respond right-click
   prompt   Insertion point
   enter   **0,0,140** <R>
   prompt   X scale ... and enter 1 <R>
   prompt   Y scale ... and enter 1 <R>
   prompt   Rotation ... and enter 0 <R>

9 Make layer SUPPORT current.

10 Select the Attach icon from the External Reference toolbar and:
   prompt   Select file to attach dialogue box
   respond a) pick r13model directory
              b) pick support.dwg file
              c) pick OK
   prompt   Attach Xref SUPPORT: SUPPORT.DWG
              SUPPORT loaded
              Insertion point
   enter   **10,0,30** <R>
   prompt   X scale ... and enter 1 <R>
   prompt   Y scale ... and enter 1 <R>
   prompt   Rotation ... and enter 0 <R>

11 Activate the layer control dialogue box and note the SUPPORT| ... layers.

12 From the menu bar select **Construct–Union** and pick the two trays and the support and:
   prompt   At least 2 solids or coplanar regions must be selected

13 What does this mean? Before inserted blocks/external references can be unioned/subtracted, they must first of all be:
   a) exploded if inserted blocks
   b) bound if external references.

14 Using the Explode icon, pick the two green trays then right-click.

15 At the command line enter **XREF** <R> and:
   prompt   ?/Bind/ ...
   enter   **B** <R> – the bind option
   prompt   Xref(s) to bind
   enter   **SUPPORT** <R>
   prompt   Scanning...

16 Check the layer control dialogue box and note that the SUPPORT|... layers have now been renamed as SUPPORT 0... .

17 Make layer MODEL current.

18 Explode the blue support.

19 Select the Interfere icon from the Solids toolbar and:
   prompt   Select the first set of solids
   respond pick the top green box then right-click
   prompt   Select the second set of solids
   respond pick the blue support then right-click
   prompt   Comparing 1 solid against 1 solid
              Interfering solids (first set): 1

(second set): 1
Interfering pairs: 1
Create interference solids?<N>
enter **Y** <R>

20 The model will be displayed with a red cylinder where interference occurs, i.e. between the top of the support and the tray. This was because the top tray was deliberately positioned so that interference would result.

21 Proceed immediately to the next section **WITHOUT** saving.

Insert exercise 2

1 Open the drawing \R13MODEL\ASSEMBLY of the tray saved earlier in the exercise **but pick NO to save changes**.

2 Lower right viewport active, layer TRAY current and UCS BASE.

3 Using INSERT at the command line:
a) insert the TRAY at 0,0,150 – full size with no rotation
b) insert the SUPPORT at 10,0,30 – full size with no rotation.

3 Explode the three inserted blocks.

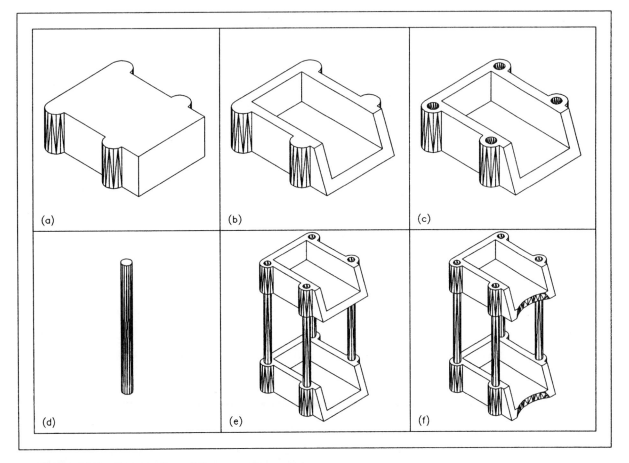

Figure 34.2 Construction of the assembly model.

228 *Modelling with AutoCAD*

4 Rectangular array the blue support:
   a) for two rows and two columns
   b) row distance: 80
   c) column distance: 70.

5 Union the six solids – fig. (e).

   Note the colour of the supports – why?

6 Create a cylinder:
   centre: 150,40,0
   radius: 50
   height: 200

7 Subtract the cylinder from the composite – fig. (f).

8 The model is complete and can be saved as ASSEMBLY.

Note

In this exercise one of the blocks was attached as an external reference, and two different layers were 'created', these being:

a) SUPPORT| ...: attached items
b) SUPPORT 0...: bound items.

Chapter **35**

Dynamic viewing solid models

The 3D dynamic view command has already been used with other 3D models, but the command is particularly useful with solid models as it allows the model to be 'cut-away' so that the user can 'see inside' the component. We will demonstrate the command with a previously created composite.

1 Open the drawing **NUCLEAR** created in Chapter 32 and refer to Fig. 35.1.

2 With the lower right viewport active restore UCS BASE with the layer MODEL current.

3 The model represents a nuclear reactor and we want to 'see' into the core.

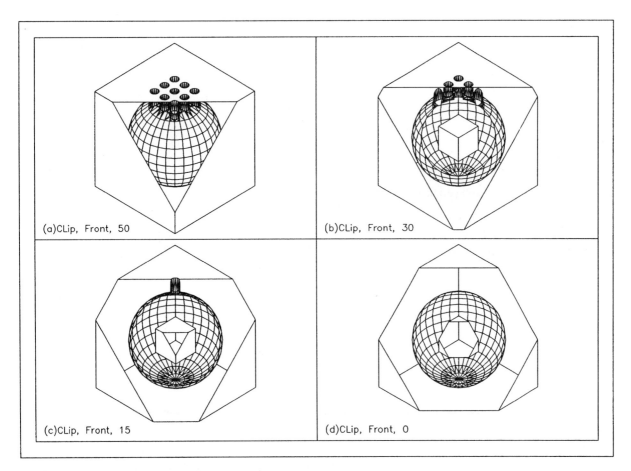

Figure 35.1 Dynamic view with NUCLEAR solid model drawing.

230 *Modelling with AutoCAD*

4 From the menu bar select **View**

3D Dynamic View

prompt	Select objects
respond	window the model then right-click
prompt	*** Switching to the WCS ***
then	CAmera ...
enter	**CL** <R>
prompt	Back/Front/<Off>
enter	**F** <R>
prompt	Eye/ON/OFF/<Distance ...
enter	**50** <R>
prompt	CAmera ...
enter	**H** <R> – fig. (a)

5 Enter the following keyboard sequence:

U <R> – undo hide
U <R> – undo clip
CL <R>
F <R>
30 <R>
H <R> – fig. (b)
U <R> – undo hide
U <R> – undo clip
CL <R>
F <R>
15 <R>
H <R> – fig. (c)
U <R> – undo hide
U <R> – undo clip
CL <R>
F <R>
0 <R>
H <R> – fig. (d)
X <R> – exit the command

6 The model will be displayed with this last dynamic view entry and HIDE and SHADE can be used

7 *Note*: this has been a very simple example to show how the dynamic view command is very useful with solid models.

Chapter 36
A solid model house

In this final exercise we will create a simple house from solid primitives using our knowledge of the R13 commands. The exercise should prove interesting at this stage and I hope that you enjoy creating the model.

The walls

1. Open your SOLA3 standard sheet with:
   a) lower right viewport active
   b) layer MODEL current
   c) UCS BASE
   d) Zoom centre about the point 150,100,100 with 500 magnification in the 3D viewport and 300 in the other three viewports.
2. Refer to Fig. 36.1.

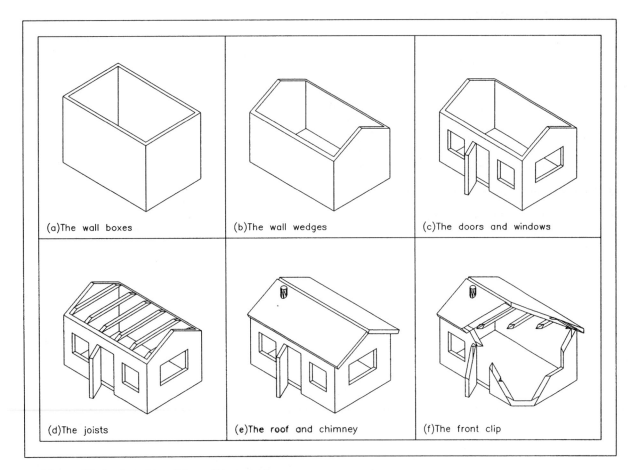

Figure 36.1 Creation of the solid model house.

3 Create two boxes with:

	Box1	Box2
corner:	0,0,0	10,10,10
length:	300	280
width :	200	180
height:	200	200
colour:	yellow	green

4 Subtract the green box from the yellow box – fig. (a).

5 Realign the UCS with the sequence:

UCS Z 90 UCS S POS1

6 Create two wedges with:

	Wedge1	Wedge2
corner:	200,0,200	0,0,200
length:	–100	100
width :	–300	–300
height:	–50	–50
colour:	yellow	yellow

7 Subtract the two wedges from the composite – fig. (b).

The windows and door

1 Restore UCS BASE.

2 Create four boxes with:

	Box1	Box2	Box3	Box4
corner:	120,0,0	30,0,50	210,0,50	300,50,50
length:	60	60	60	–10
width :	10	10	10	100
height:	120	60	60	60
colour:	red	red	red	red

3 Subtract the boxes from the composite.

4 Create the door from a box with:
corner:	120,0,0
length:	60
width :	10
height:	120
colour:	green

5 2D rotate this door using the last option:
a) about the point 120,0
b) with a rotation of –65
c) fig. (c).

6 Try hide and shade at this stage then regenall

The roof joists

Roof joists will be created from an extruded polyline which has its own UCS setting, so:

1 At the command line enter **UCS** <R> then **3** <R> and enter:
a) origin point: 290,10,155
b) X-axis point: 290,190,155
c) Y-axis point: 290,10,200.

2 Save this UCS as JOIST.

3 Make a new layer named JOIST with colour blue and current.

4 Using the polyline icon draw:
From   0,0
To   @20<26.57
To   @144.22<0
To   @20<−26.57
To   close

5 Using the Extrude icon:
a) pick the blue polyline
b) enter −10 as the height
c) enter 0 as the taper angle.

6 Restore UCS BASE.

7 Rectangular array the blue extrusion:
a) for one row and six columns
b) with a column distance of −54.

8 Result of these operations is fig. (d).

9 **Do not union** the joists with the composite

10 *Questions*
a) What is the 26.57 angle for?
b) How did I work out the −54 array distance?

The roof

The roof will be created on its own layer with its own UCS so:

1 UCS at BASE.

2 Make a new layer named ROOF, colour magenta and current.

3 Set a new UCS using the three-point option with:
a) origin point: 10,10,155
b) X-axis point: 300,10,155
c) Y-axis point: 10,100,200
d) save as ROOF.

4 Create two boxes with:

	Box1	*Box2*
corner:	−20,100.62,0	−20,100.62,0
length:	320	320
width :	−120	120
height:	10	10

234 *Modelling with AutoCAD*

5 Using **Construct–3D Rotate**:
   a) pick box2 then right-click
   b) enter *X* as the axis
   c) enter –10,100.62,0 as a point an the axis
   d) enter –53.14 as the rotation angle – why this?

6 Restore UCS BASE and create a chimney from a cylinder with:
   centre: 50,70,190
   radius: 8
   height: 30

7 Union the two roof parts and the chimney – fig. (e).

8 Hide and shade and no joists will be seen.

9 Save the model at this stage as HOUSE as it is complete. I have displayed the four viewport configuration of the house in Fig. 36.2. The roof layer has been removed for obvious reasons?

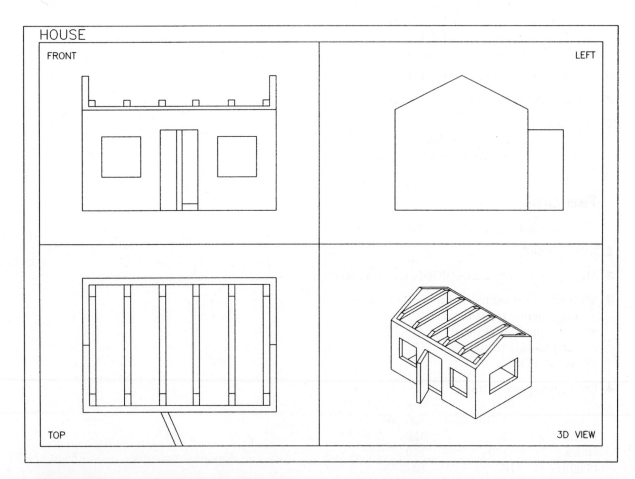

Figure 36.2 The completed house solid model.

Clipping the model

1 UCS BASE and layer MODEL current.

2 At the command line enter **DVIEW** <R> and:
   a) window the complete house then right-click
   b) enter CL <R> – the clip option
   c) enter F <R> – for a front clip
   d) enter 30 <R> – the clip distance
   e) enter X <R> – to exit the command.

3 The result of this operation is fig. (f).

4 Hide and shade – interesting result.

5 This completes the last exercise.

Chapter **37**

Finally

By the time that you read this, I hope that you have managed to complete the various exercises in the previous chapters. If you have, you now have the necessary under-standing of modelling with AutoCAD Release 13. The progression has been from extruded drawings ($2^1/_2$D) through wire-frame model, surface models to solid models. My exercises have tended to be engineering oriented, but the principles are the same whatever the CAD discipline.

The object of this book has been to teach the reader modelling with Release 13, and I hope that you have enjoyed reading the book as much as I have had in preparing it. Any comments you have to make would be more than appreciated.

Bob McFarlane

Tutorials

Tutorial 1: Using the reference sizes given, construct a 3D extruded model of the half-coupling. View your model at different viewpoints.

REFERENCE SIZES

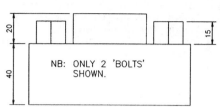
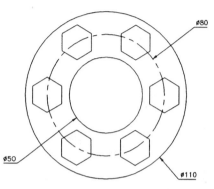

(a) 3D Viewpoint Preset SE Isometric

(b) SE Isometric with HIDE

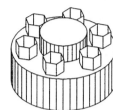

(c) Command line VPOINT 'R' 315°, −5°

(d) Command line VPOINT with HIDE.

NB: ONLY 2 'BOLTS' SHOWN.

NOTE: The 'bolts' have been drawn with the POLYGON command, 6 sides circumscribed on a radius 10 circle, then polar array for 6 items with rotation.

Tutorial 2: Draw a wire-frame model of the MILL GUIDE block and add all text.

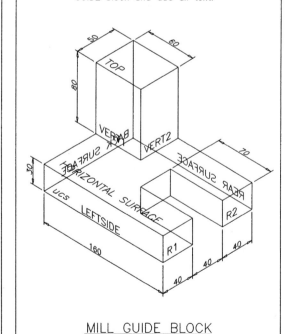

MILL GUIDE BLOCK

Tutorial 3: Draw the SPECIAL SLIP BLOCK as a wire-frame model and add all text.

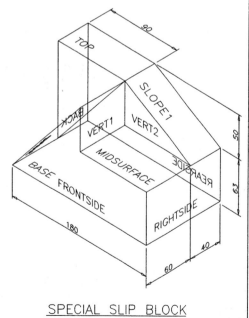

SPECIAL SLIP BLOCK

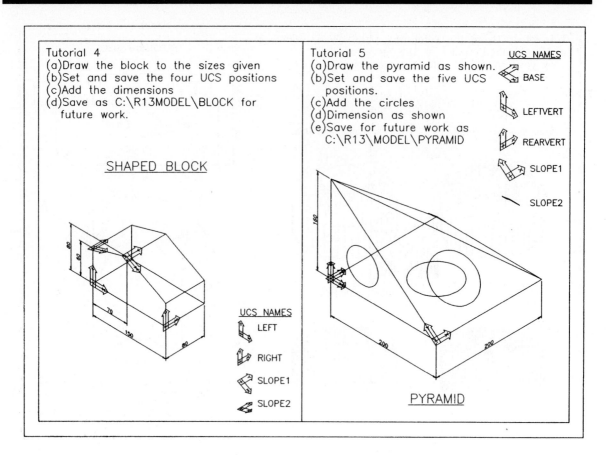
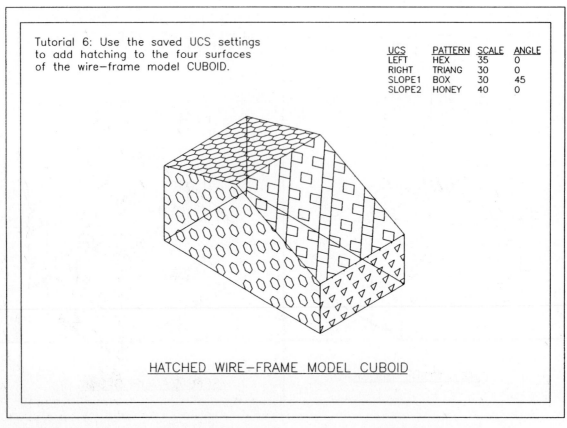

Tutorial 7: Make five new layers, then use the saved UCS positions to add hatching to the faces of the wire-frame model PYRAMID.

UCS	PATTERN	SCALE,ANGLE
BASE	AR-PARQ1	3,0
LEFTVERT	AR-B816	1,0
RIGHTVERT	AR-B816	1,0
SLOPE1	AR-B816	1,0
SLOPE2	AR-B816	1,0

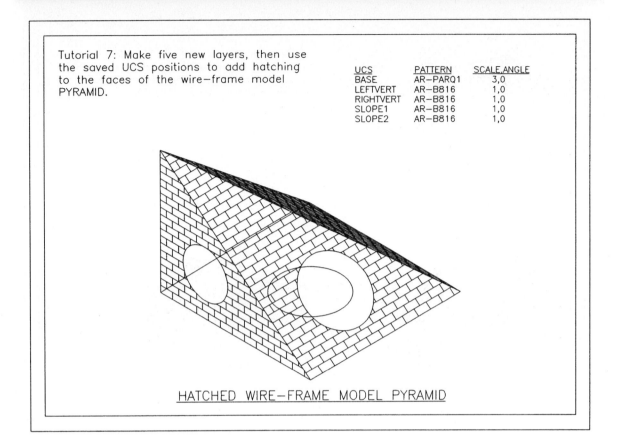

HATCHED WIRE-FRAME MODEL PYRAMID

Tutorial 8
Set the three viewport configuration for the special slip block created in Tutorial 3.

SPECIAL SLIP BLOCK

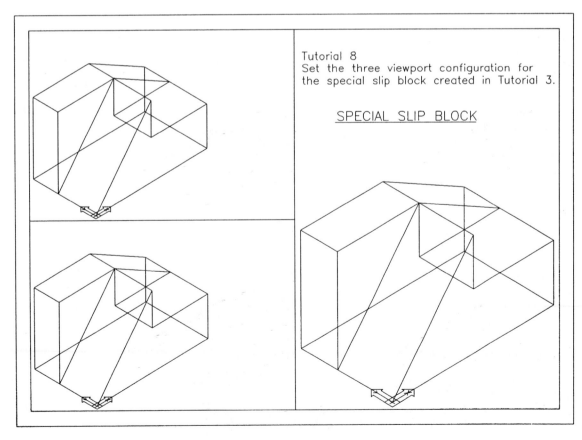

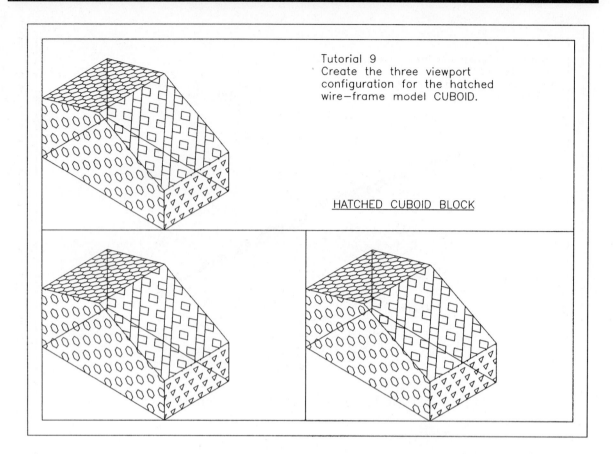

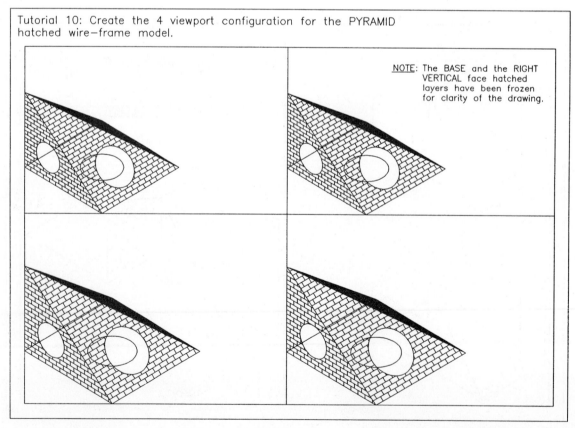

Tutorials **241**

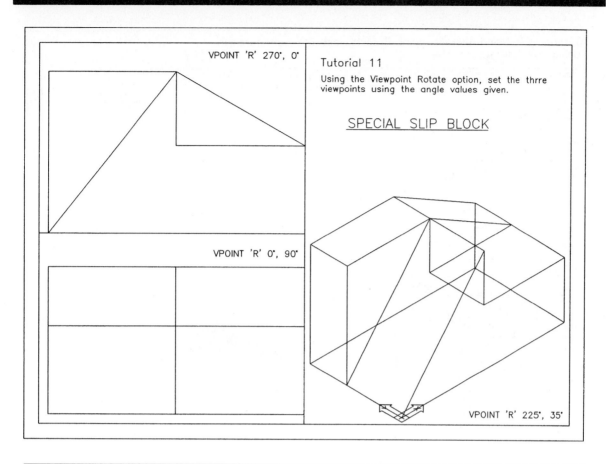

VPOINT 'R' 270°, 0°

VPOINT 'R' 0°, 90°

VPOINT 'R' 225°, 35°

Tutorial 11
Using the Viewpoint Rotate option, set the thrre viewpoints using the angle values given.

SPECIAL SLIP BLOCK

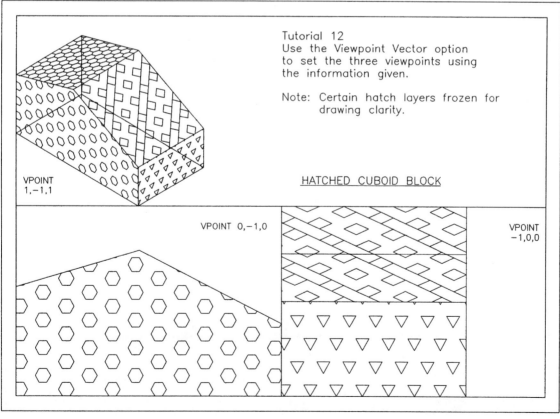

Tutorial 12
Use the Viewpoint Vector option to set the three viewpoints using the information given.

Note: Certain hatch layers frozen for drawing clarity.

HATCHED CUBOID BLOCK

VPOINT 1,−1,1

VPOINT 0,−1,0

VPOINT −1,0,0

Tutorial 13: Using the Viewpoint Presets, set the configuration using the information given.

FRONT LEFT

TOP SW ISOMETRIC

NOTE
Certain layers have bee frozen for drawing clarity.

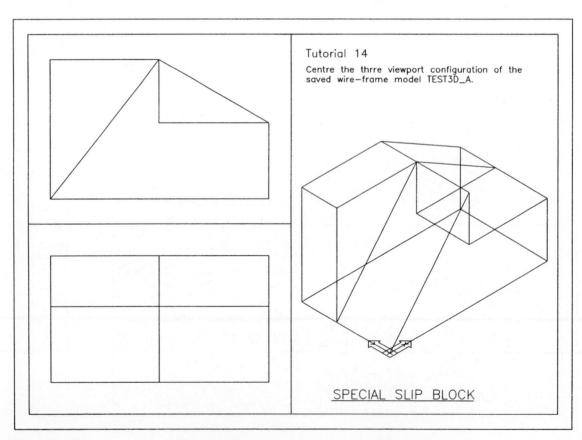

Tutorial 14
Centre the thrre viewport configuration of the saved wire-frame model TEST3D_A.

SPECIAL SLIP BLOCK

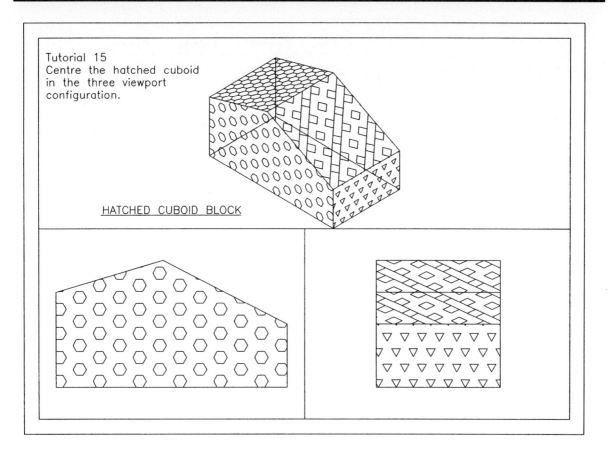

Tutorial 15
Centre the hatched cuboid in the three viewport configuration.

HATCHED CUBOID BLOCK

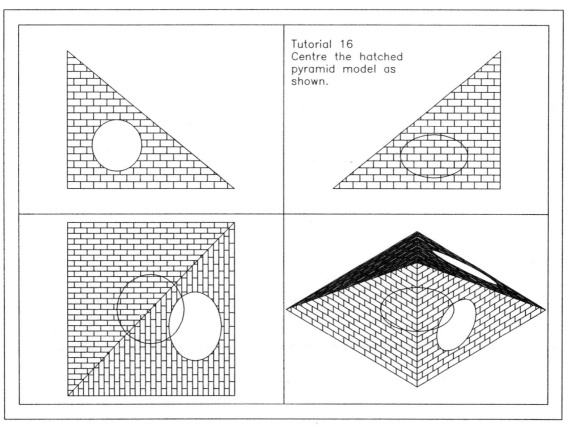

Tutorial 16
Centre the hatched pyramid model as shown.

244 Modelling with AutoCAD

Tutorial 17
Draw the wire-frame model of the cheese to the sizes given. Using the 3DFACE command, add faces to each surface, then display the model in a four viewport configuration.

Tutorial 18
Set a 4 viewport configuration with the given viewpoints. Draw the wire-frame model using the sizes given.
Convert the wire-frame model to a surface model using the Ruked Surface command.
Hide and shade.

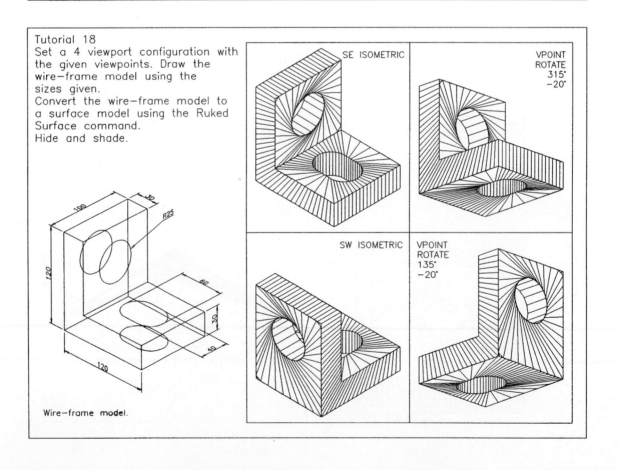

Tutorials **245**

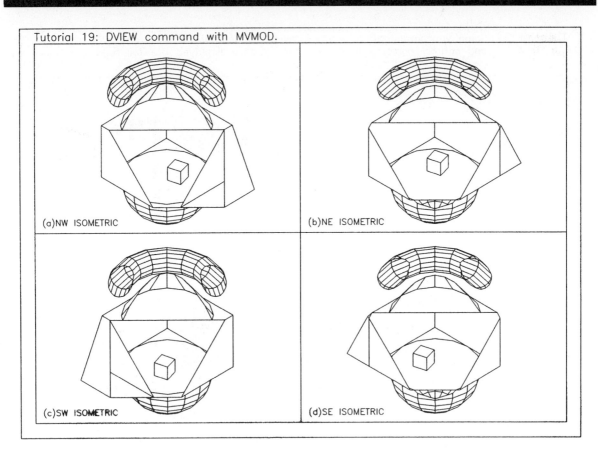

Tutorial 19: DVIEW command with MVMOD.
(a) NW ISOMETRIC
(b) NE ISOMETRIC
(c) SW ISOMETRIC
(d) SE ISOMETRIC

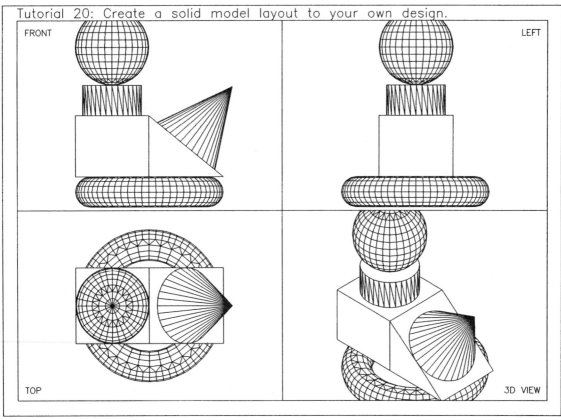

Tutorial 20: Create a solid model layout to your own design.
FRONT
LEFT
TOP
3D VIEW

246 *Modelling with AutoCAD*

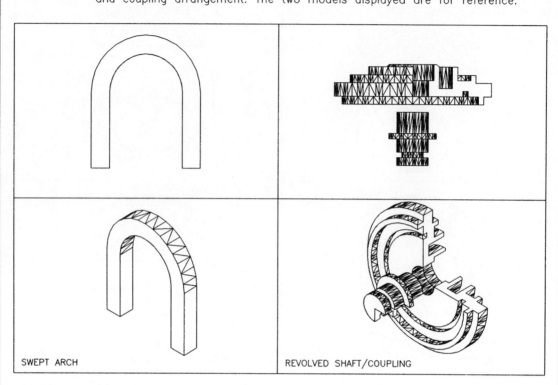

Tutorial 21: Using your imagination produce swept models of an arch and a shaft and coupling arrangement. The two models displayed are for reference.

SWEPT ARCH

REVOLVED SHAFT/COUPLING

Index

3D Array 125, 129, 131, 224
3D Coordinate 14, 17
3D Dynamic view 132, 139, 218, 229, 230
3D Face 74, 75, 76, 77
3D Mesh 86, 87
3D Mirror 127, 131, 215
3D Objects 40, 41, 54
3D Poly 83, 84
3D Rotate 125, 126, 131, 234
3D Viewpoint 60, 61, 62, 64
3D Viewpoint presets 4, 7, 8, 13, 15, 19, 31, 51,
    53, 54, 64, 115, 218

Align 130, 131, 202
Ambiguity 4, 67, 72, 145, 224
Array 7, 35, 208
Axis of revolution 98, 100

BASE 224, 225
BHATCH 212
Boolean 146, 172, 178
Boundary hatch 44
BOX primitive 151, 152
BREP 146
Bull's-eye 63

CAmera 132, 134
Chamfer 20, 191, 194
CHPROP 152
CLip 136, 235
Closed path 89, 93
COLOR 74
Composites 173, 174, 178, 188, 193, 205, 213
CONE primitive 157
COONS patch s107
CREP 146
CSG 146
Cylinder primitive 155, 156

Dialogue boxes
    Boundary Creation 200
    Boundary Definition 44
    Object Creation 2
    Tiled Viewports 49
    UCS 29
    UCS Presets 61
    View Control 66
    Viewpoint Presets 61

Dimension
    geometry 37
    format 37
    annotation 37
Direction vector 94, 95, 96
Distance 137
DVIEW 139, 235

Edge 78
Edit Polyline 84, 86, 105, 106, 164, 166, 185
EDGESURF 103
Edge surfaces 102, 103, 104
Edit text 121
ELEV 2, 3, 7, 9
Elevation 1
External reference 223, 228
Extruded solids 163, 166, 167, 168, 196

FACETED 72
FACETRES 150, 156, 160
Fillet 7, 191, 194
Floating viewports 48, 110, 111, 113, 115, 117,
    122, 184

Global layers 141

Hide 5, 8, 9, 46, 74, 153

Icons
    3D Array 129
    3D Face 75
    3D Mirror 128
    3D Rotate 125
    Attack 226
    Boundary 199
    Box 152
    Cone 158
    Cylinder 156
    Edge surface 103
    Extruded solid 163
    Interfere 226
    Intersection 178
    Mass property 176
    Pyramid 41
    Region 196
    Revolved solid 171
    Revolved surface 98
    Ruled surface 89

Icons (*continued*)
  Section 211
  Slice 207
  Sphere 160
  Subtraction 180
  Torus 160
  Tabulated surface 95
  Union 178
  Wedge 154
  World 13
Insert 141, 225
Interference 223, 227
Intersection 174, 215
Isolines 149, 150, 185, 192
Invisible 3D edge 78, 85

Loops 196

MASSPROP 187
Multiple viewports 108, 116, 119

Named UCS 29

Object creation 4
Open path 89, 93

PAM 137
Path curve 94, 95, 96, 98, 106
PFACE 80, 81
PLAN 30, 31
Plan view 8, 218
POints 136
Polygon mesh 101

Realign viewports 221
REDRAW 54
REGEN 54, , 76, 91
Regions 196, 198, 199
Revolved solids 170, 172, 198
Revolved surfaces 100
Ruled surfaces 88, 89

Section solids 210, 211, 212, 220
Set UCS 52
Shade 8, 75, 89, 153
Solid assembly 223
Solid modeller 145, 146
Solid modelling 145, 146
Slicing solids 205, 206, 209, 212
Sphere primitive 159
Spline 84
Subtraction 174, 176

Surfaces 40, 41, 54, 72, 75, 78, 89, 95, 103
Surface modelling 145, 146
Surftab 89, 90, 92, 93, 97, 98, 101, 102,
    103, 104, 114
SURFTYPE 107

Tabulated surface 95
Target 132, 134
Thickness 1
TILEMODE 49, 56, 108, 114
Tiled viewports 48, 50, 51, 52, 68
Toggle model/paper space 109
Toolbars
  surface 73
  UCS 28
Topology 146
Torus primitive 160
True shape 220, 221
Twist 134

UCS 10, 11, 13, 15, 16, 17, 18, 22, 23, 24,
    25, 36, 38, 39, 64, 67, 208
UCSFOLLOW 32
UCS icon 11, 12, 32
UCSICON 119, 178, 216
UCS options 26
Union 174, 176, 226
Untitled viewports 48, 114

View 4, 49, 52, 55, 68, 109, 110, 111, 115
VIEW 66
Viewpoint
  rotate 58, 60
  tripod 62
  vector 64
Viewport layer control 142, 144, 180, 219
Viewport specific layers 142, 219
VPOINT 5, 9, 58, 59, 64
VPLAYER 144
VPORTS 51

WCS 10, 17, 64, 67
Wedge primitive 153, 154
Wireframe 18, 91, 97, 145

XREF 226

Zoom 137
Zoom centre 55, 56, 68, 70, 115, 121, 191,
    201, 210, 214
Zoom paper space 112
Zoom XP 117, 193

City College Norwich